Cambridge Tracts in Mathematics and Mathematical Physics

GENERAL EDITOR

W. V. D. HODGE, M.A.

No. 18

The General Theory of Dirichlet's Series

THE GENERAL THEORY OF DIRICHLET'S SERIES

by

G. H. HARDY

and

MARCEL RIESZ

Cambridge:
at the University Press
1952

CAMBRIDGE
UNIVERSITY PRESS

University Printing House, Cambridge CB2 8BS, United Kingdom

Cambridge University Press is part of the University of Cambridge.

It furthers the University's mission by disseminating knowledge in the pursuit of education, learning and research at the highest international levels of excellence.

www.cambridge.org
Information on this title: www.cambridge.org/9781107493872

First published 1952
Re-issued 2015

A catalogue record for this publication is available from the British Library

ISBN 978-1-107-49387-2 Paperback

MATHEMATICIS QUOTQUOT UBIQUE SUNT
OPERUM SOCIETATEM NUNC DIREMPTAM
MOX UT OPTARE LICET REDINTEGRATURIS
D.D.D. AUCTORES
HOSTES IDEMQUE AMICI

PREFACE

THE publication of this tract has been delayed by a variety of causes, and I am now compelled to issue it without Dr Riesz's help in the final correction of the proofs. This has at any rate one advantage, that it gives me the opportunity of saying how conscious I am that whatever value it possesses is due mainly to his contributions to it, and in particular to the fact that it contains the first systematic account of his beautiful theory of the summation of series by 'typical means'.

The task of condensing any account of so extensive a theory into the compass of one of these tracts has proved an exceedingly difficult one. Many important theorems are stated without proof, and many details are left to the reader. I believe, however, that our account is full enough to serve as a guide to other mathematicians researching in this and allied subjects. Such readers will be familiar with Landau's *Handbuch der Lehre von der Verteilung der Primzahlen*, and will hardly need to be told how much we, in common with all other investigators in this field, owe to the writings and to the personal encouragement of its author.

G. H. H.

19 *May* 1915.

CONTENTS

THE GENERAL THEORY OF DIRICHLET'S SERIES

I

INTRODUCTION

1. The series whose theory forms the subject of this tract are of the form

$$f(s) = \sum_{1}^{\infty} a_n e^{-\lambda_n s} \quad \text{.........................(1)},$$

where (λ_n) is a sequence of real increasing numbers whose limit is infinity, and $s = \sigma + ti$ is a complex variable whose real and imaginary parts are σ and t. Such a series is called a Dirichlet's series of *type* λ_n. If $\lambda_n = n$, then $f(s)$ is a power series in e^{-s}. If $\lambda_n = \log n$, then

$$f(s) = \sum_{1}^{\infty} a_n n^{-s} \quad \text{.........................(2)}$$

is called an *ordinary* Dirichlet's series.

Dirichlet's series were, as their name implies, first introduced into analysis by Dirichlet, primarily with a view to applications in the theory of numbers. A number of important theorems concerning them were proved by Dedekind, and incorporated by him in his later editions of Dirichlet's *Vorlesungen über Zahlentheorie.* Dirichlet and Dedekind, however, considered only real values of the variable s. The first theorems involving complex values of s are due to Jensen*, who determined the nature of the region of convergence of the general series (1); and the first attempt to construct a systematic theory of the function $f(s)$ was made by Cahen† in a memoir which, although much of the analysis which it contains is open to serious criticism, has

* Jensen, **1**, **2**. References in thick type are to the bibliography at the end of the tract.

† Cahen, **1**.

served—and possibly just for that reason—as the starting point of most of the later researches in the subject*.

It is clear that all but a finite number of the numbers λ_n must be positive. It is often convenient to suppose that they are all positive, or at any rate that $\lambda_1 \geqq 0$.†

2. It will be convenient at this point to fix certain notations which we shall regard as stereotyped throughout the tract.

(i) By $[x]$ we mean the algebraically greatest integer not greater than x. By

$$\sum_{\alpha}^{\beta} f(n)$$

we mean the sum of all values of $f(n)$ for which $\alpha \leqq n \leqq \beta$, *i.e.* for $[\alpha] \leqq n \leqq [\beta]$ or $[\alpha] < n \leqq [\beta]$, according as α is or is not an integer. We shall also write

$$A(x) = \sum_{1}^{x} a_n, \quad A(x, y) = \sum_{x}^{y} a_n, \ddagger$$

$$\Delta a_n = a_n - a_{n+1}.$$

(ii) We shall follow Landau in his use of the symbols o, O.§ That is to say, if ϕ is a positive function of a variable which tends to a limit, we shall write

$$f = o(\phi)$$

if $f/\phi \to 0$, and

$$f = O(\phi)$$

if $|f|/\phi$ remains less than a constant K. We shall use the letter K to denote an unspecified constant, not always the same||.

* Fuller information as to the history of the subject (up to 1909) will be found in Landau's *Handbuch der Lehre von der Verteilung der Primzahlen*, Vol. 2, Book 6, Notes and Bibliography, and in the *Encycl. des sc. math.*, T. 1, Vol. 3, pp. 249 *et seq.* We shall refer to Landau's book by the letter *H*. The two volumes are paged consecutively.

† It is evident that we can reduce the series (1) to a series satisfying this condition either (*a*) by subtracting from $f(s)$ a finite sum $\Sigma a_n e^{-\lambda_n s}$ or (*b*) by multiplying $f(s)$ by an exponential e^{-Cs}. These operations would of course change the type of the series.

‡ We shall use the corresponding notations, with letters other than a, without further explanation.

§ Landau, *H.*, p. 883, states that the symbol O seems to have been first used by Bachmann, *Analytische Zahlentheorie*, Vol. 2, p. 401.

|| For fuller explanations see Hardy, *Orders of infinity* (Camb. Math. Tracts, No. 12), pp. 5 *et seq.*

II

ELEMENTARY THEORY OF THE CONVERGENCE OF DIRICHLET'S SERIES

1. Two fundamental lemmas. Much of our argument will be based upon the two lemmas which follow.

LEMMA 1. *We have identically*

$$\sum_x^y a_\nu \phi(\nu) = \sum_x^{y-1} A(x, n) \Delta \phi(n) + A(x, y) \phi[y].$$

This is Abel's classical lemma on partial summation*.

LEMMA 2. *If* $\sigma \neq 0$, *then*

$$|\Delta e^{-\lambda_n s}| \leq \frac{|s|}{\sigma} \Delta e^{-\lambda_n \sigma}. \dagger$$

For

$$|\Delta e^{-\lambda_n s}| = \left| \int_{\lambda_n}^{\lambda_{n+1}} s e^{-us} \, du \right| \leq |s| \int_{\lambda_n}^{\lambda_{n+1}} e^{-u\sigma} \, du = \frac{|s|}{\sigma} \Delta e^{-\lambda_n \sigma}.$$

2. Fundamental Theorems. Region of convergence, analytical character, and uniqueness of the series. We are now in a position to prove the most important theorems in the elementary theory of Dirichlet's series.

THEOREM 1. *If the series is convergent for* $s = \sigma + ti$, *then it is convergent for any value of* s *whose real part is greater than* σ.

This theorem is included in the more general and less elementary theorem which follows.

THEOREM 2. *If the series is convergent for* $s = s_0$, *then it is uniformly convergent throughout the angular region in the plane of* s *defined by the inequality*

$$|\text{am}\,(s - s_0)| \leq \alpha < \tfrac{1}{2}\pi. \ddagger$$

* Abel, **1**.

† This lemma seems to have been stated first in this form by Perron, **1**, but is contained implicitly in many earlier writings.

‡ If $s = re^{i\theta}$, we write $r = |s|$, $\theta = \text{am}\, s$. Theorem 1 is due to Jensen, **1**, and Theorem 2 to Cahen, **1**.

We may clearly suppose $s_0 = 0$ without loss of generality. We have

$$\sum_m^n a_\nu e^{-\lambda_\nu s} = \sum_m^{n-1} A(m, \nu) \Delta e^{-\lambda_\nu s} + A(m, n) e^{-\lambda_n s},$$

by Lemma 1. If ϵ is assigned we can choose m_0 so that $\lambda_m > 0$ and

$$|A(m, \nu)| < \epsilon \cos \alpha$$

for $\nu \geqq m \geqq m_0$. If now we apply Lemma 2, and observe that

$$|s|/\sigma \leqq \sec \alpha$$

throughout the region which we are considering, we obtain

$$\left|\sum_m^n a_\nu e^{-\lambda_\nu s}\right| < \epsilon \left(\sum_m^{n-1} \Delta e^{-\lambda_\nu \sigma} + e^{-\lambda_n \sigma}\right) = \epsilon e^{-\lambda_m \sigma} < \epsilon$$

for $n \geqq m \geqq m_0$. Thus Theorem 2 is proved*, and Theorem 1 is an obvious corollary.

There are now three possibilities as regards the convergence of the series. It may converge for *all*, or *no*, or *some* values of s. In the last case it follows from Theorem 1, by a classical argument, that we can find a number σ_0 such that the series is convergent for $\sigma > \sigma_0$ and divergent or oscillatory for $\sigma < \sigma_0$.

THEOREM 3. *The series may be convergent for all values of s, or for none, or for some only. In the last case there is a number σ_0 such that the series is convergent for $\sigma > \sigma_0$ and divergent or oscillatory for $\sigma < \sigma_0$.*

In other words *the region of convergence is a half-plane*†. We shall call σ_0 the *abscissa of convergence*, and the line $\sigma = \sigma_0$ the *line of convergence*. It is convenient to write $\sigma_0 = -\infty$ or $\sigma_0 = \infty$ when the series is convergent for all or no values of s. *On* the line of convergence the question of the convergence of the series remains open, and requires considerations of a much more delicate character.

* It is possible to substitute for the angle considered in this theorem a wider region; *e.g.* the region

$$\sigma \geqq 0, \quad |t| \leqq e^{K\sigma} - 1$$

(Perron, **1**; Landau, *H.*, p. 739). We shall not require any wider theorem than 2. It may be added that the result of Theorem 1 remains true when we only assume that Σa_n is at most finitely oscillating: in fact, with this hypothesis, the result of Theorem 2 holds for the region

$$|\text{am}(s - s_0)| \leqq \alpha < \tfrac{1}{2}\pi, \quad \sigma \geqq \delta > 0,$$

as is easily proved by a trifling modification of the argument given above.

† Jensen, **1**.

3. Examples. (i) The series $\sum a^n n^{-s}$, where $|a| < 1$, is convergent for all values of s.

(ii) The series $\sum a^n n^{-s}$, where $|a| > 1$, is convergent for no values of s.

(iii) The series $\sum n^{-s}$ has $\sigma = 1$ as its line of convergence. It is not convergent at any point of the line of convergence, diverging to $+\infty$ for $s=1$, and oscillating finitely* at all other points of the line.

(iv) The series $\sum_2^\infty (\log n)^{-2} n^{-s}$ has the same line of convergence as the last series, but is convergent (indeed absolutely convergent) at all points of the line.

(v) The series $\sum_2^\infty a_n n^{-s}$, where $a_n = (-1)^n + (\log n)^{-2}$, has the same line of convergence, and is convergent (though not absolutely) at all points of it†.

4. THEOREM 4. *Let D denote any finite region in the plane of s for all points of which*

$$\sigma \geqq \sigma_0 + \delta > \sigma_0.$$

Then the series is uniformly convergent throughout D, and its sum $f(s)$ is a branch of an analytic function, regular throughout D. Further, the series

$$\sum a_n \lambda_n^{\rho} e^{-\lambda_n s},$$

where ρ is any number real or complex, and λ_n^{ρ} has its principal value, is also uniformly convergent in D, and, when ρ is a positive integer, represents the function

$$(-1)^\rho f^{(\rho)}(s).$$

The uniform convergence of the original series follows at once from Theorem 2, since we can draw an angle of the type considered in that theorem and including D‡. The remaining results, in so far as they concern the original series and its derived series, then follow immediately from classical theorems of Weierstrass§.

When ρ is not a positive integer, we choose a positive integer m so that the real part of $\rho - m$ is negative. The series

$$\sum a_n \lambda_n^{\rho - m} e^{-\lambda_n s} \quad \ldots\ldots\ldots\ldots (1)$$

may be written in the form

$$\sum b_n e^{-(m-\rho)\log \lambda_n} \quad \ldots\ldots\ldots\ldots (2),$$

where $b_n = a_n e^{-\lambda_n s}$. Regarding (2) as a Dirichlet's series of type $\log \lambda_n$, and applying Theorem 1, we see that (1) is convergent whenever

* See, *e.g.*, Bromwich, **2**.

† We are indebted to Dr Bohr for this example.

‡ The vertex of the angle may be taken at σ_0, if the series is convergent for $s = \sigma_0$, and otherwise at $\sigma_0 + \eta$, where $0 < \eta < \delta$.

§ See, *e.g.*, Weierstrass, *Abhandlungen aus der Funktionentheorie*, pp. 72 *et seq.*; Osgood, *Funktionentheorie*, Vol. 1, pp. 257 *et seq.*

$\Sigma a_n e^{-\lambda_n s}$ is convergent. The proof of the theorem may now be completed by a repetition of our previous arguments.

THEOREM 5. *If the series is convergent for $s = s_0$, and has the sum $f(s_0)$, then $f(s) \to f(s_0)$ when $s \to s_0$ along any path which lies entirely within the region*

$$|\operatorname{am}(s - s_0)| \leqq \alpha < \tfrac{1}{2}\pi.$$

This theorem* is an immediate corollary from Theorem 2. It is of course only when s_0 lies on the line of convergence that it gives us any information beyond what is given by Theorem 4.

5. THEOREM 6. *Suppose that the series is convergent for $s = 0$, and let E denote the region*

$$\sigma \geqq \delta > 0, \quad |\operatorname{am} s| \leqq \alpha < \tfrac{1}{2}\pi.$$

Suppose further that $f(s) = 0$ for an infinity of values of s lying inside E. Then $a_n = 0$ for all values of n.

The function $f(s)$ cannot have an infinity of zeros in the neighbourhood of any finite point of E, since it is regular at any such point. Hence we can find an infinity of values $s_n = \sigma_n + t_n i$, where $\sigma_{n+1} > \sigma_n$, $\lim \sigma_n = \infty$, such that $f(s_n) = 0$.

But
$$g(s) = e^{\lambda_1 s} f(s) = a_1 + \sum_2^\infty a_n e^{-(\lambda_n - \lambda_1) s}$$

is convergent for $s = 0$ and so uniformly convergent in E. Hence

$$g(s) \to a_1$$

when $s \to \infty$ along any path in E. This contradicts the fact that $g(s_n) = 0$, unless $a_1 = 0$. It is evident that we may repeat this argument and so complete the proof of the theorem †.

6. Determination of the abscissa of convergence. Let us suppose that the series is not convergent for $s = 0$, and let

$$\overline{\lim} \frac{\log |A(n)|}{\lambda_n} = \gamma. \ddagger$$

* The generalisation of the 'Abel-Stolz' theorem for power series (Abel, **1**; Stolz, **1**, **2**).

† This theorem, like Theorem 2 itself, may be made wider: see Perron, **1**; Landau, *H.*, p. 745. Until recently it was an open question whether it were possible that $f(s)$ could have zeros whose real parts surpass all limit: all that Theorem 6 and its generalisations assert is that the imaginary parts of such zeros, if they exist, must increase with more than a certain rapidity. The question has however been answered affirmatively by Bohr, **4**. But if there is a region of absolute convergence, the answer is negative (see III, § 5).

‡ By $\overline{\lim}\, u_n$ we denote the 'maximum limit' of the sequence u_n: *cf.* Bromwich, *Infinite series*, p. 13.

It is evident that $\gamma \geqq 0$*. We shall now prove that $\sigma_0 = \gamma$.

(i) Let δ be any positive number. We shall prove first that the series is convergent for $s = \gamma + \delta$.

Choose ϵ so that $0 < \epsilon < \delta$. Then, by the definition of γ, we have

$$\log |A(\nu)| < (\gamma + \delta - \epsilon)\lambda_\nu, \quad |A(\nu)| < e^{(\gamma+\delta-\epsilon)\lambda_\nu}$$

for sufficiently large values of ν. Now

$$\sum_1^n a_\nu e^{-\lambda_\nu s} = \sum_1^{n-1} A(\nu)\,\Delta e^{-\lambda_\nu s} + A(n)\,e^{-\lambda_n s}.$$

The last term is, for sufficiently large values of n, less in absolute value than $e^{-\epsilon\lambda_n}$, and so tends to zero; and everything depends on establishing the convergence of the series

$$\Sigma\, e^{(\gamma+\delta-\epsilon)\lambda_\nu}\,\Delta\, e^{-(\gamma+\delta)\lambda_\nu}.$$

Now, since $\gamma + \delta - \epsilon$ is positive, we have

$$e^{(\gamma+\delta-\epsilon)\lambda_\nu}\Delta\, e^{-(\gamma+\delta)\lambda_\nu} = (\gamma+\delta)\int_{\lambda_\nu}^{\lambda_{\nu+1}} e^{(\gamma+\delta-\epsilon)\lambda_\nu - (\gamma+\delta)x}\,dx$$

$$< (\gamma+\delta)\int_{\lambda_\nu}^{\lambda_{\nu+1}} e^{-\epsilon x}\,dx;$$

and the series $$(\gamma+\delta)\,\Sigma\int_{\lambda_\nu}^{\lambda_{\nu+1}} e^{-\epsilon x}\,dx$$

is obviously convergent. It follows that

$$\sigma_0 \leqq \gamma.$$

(ii) Suppose $$\Sigma\, a_\nu e^{-\lambda_\nu s} = \Sigma\, b_\nu \qquad (s > 0)$$

convergent. Then

$$A(n) = \sum_1^n b_\nu e^{\lambda_\nu s} = \sum_1^{n-1} B(\nu)\,\Delta e^{\lambda_\nu s} + B(n)\,e^{\lambda_n s}.$$

It follows that $$|A(n)| < K e^{\lambda_n s},$$

and therefore that

$$\log |A(n)| < \lambda_n s + K < (s+\delta)\lambda_n,$$

for any positive δ, if n is large enough. Hence

$$s \geqq \overline{\lim}\, \frac{\log |A(n)|}{\lambda_n} = \gamma,$$

and therefore $$\sigma_0 \geqq \gamma.$$

* We can determine a constant K such that $\log |A(n)| > -K$ for an infinity of values of n. This would still be true if Σa_n converged to a sum other than zero: but if the sum were zero we should have

$$\log |A(n)| \to -\infty.$$

From the results of (i) and (ii) we deduce

THEOREM 7. *If the abscissa of convergence of the series is positive, it is given by the formula*

$$\sigma_0 = \overline{\lim} \frac{\log |A(n)|}{\lambda_n}.*$$

7. Absolute convergence of Dirichlet's series. We can apply the arguments of the preceding sections to the series

$$\Sigma |a_n| e^{-\lambda_n s} \quad \text{..............................(1).}$$

We deduce the following result:

THEOREM 8. *There is a number $\bar{\sigma}$ such that the series* (1) *is absolutely convergent if $\sigma > \bar{\sigma}$ and is not absolutely convergent if $\sigma < \bar{\sigma}$. This number, if positive, is given by the formula*

$$\bar{\sigma} = \overline{\lim} \frac{\log \bar{A}(n)}{\lambda_n},$$

where $\bar{A}(n) = |a_1| + |a_2| + \dots + |a_n|.$

In other words a Dirichlet's series possesses, besides its abscissa, line, and half-plane of convergence, an abscissa, line, and half-plane of absolute convergence. It should however be observed that the theorem which asserts the existence of a half-plane of absolute convergence is in reality more elementary than Theorem 3, as it follows at once from the inequality

$$|e^{-\lambda_\nu s}| \leq |e^{-\lambda_\nu s_1}| \qquad (\sigma \geqq \sigma_1),$$

and does not depend on Lemma 1.

It is evident that $\bar{\sigma} \geqq \sigma_0$. We may of course have $\bar{\sigma} = \infty$ or $\bar{\sigma} = -\infty$. In general there will be a *strip* between the lines of convergence and absolute convergence, throughout which the series is conditionally convergent. This strip may vanish (if $\bar{\sigma} = \sigma_0$) or comprise the whole plane (if $\sigma_0 = -\infty$, $\bar{\sigma} = \infty$) or a half-plane (if

* Cahen, **1**. Dedekind, *l.c.* p. 1, and Jensen, **2**, had already given results which together contain the substance of the theorem. The result holds when $\sigma_0 = 0$, unless Σa_n converges to zero. If $\sigma_0 < 0$ the result is in general untrue. It is plain that in such a case we can find σ_0 by first applying to the variable s such a linear transformation as will make the abscissa of convergence positive. But there is a formula directly applicable to this case, viz.

$$\sigma_0 = \overline{\lim} \frac{\log |A - A(n)|}{\lambda_{n+1}},$$

where A is the sum of the series Σa_n (obviously convergent when $\sigma_0 < 0$). This formula was given (with a slight error, viz. λ_n for λ_{n+1}) by Pincherle, **1**: see also Knopp, **6**; Schnee, **6**. Formulae applicable in *all* cases have been found by Knopp, **6** (for the case $\lambda_n = \log n$ only); Kojima, **1**; Fujiwara, **1**; and Lindh (Mittag-Leffler, **1**).

$\sigma_0 = -\infty,\ -\infty < \bar{\sigma} < \infty$ or $-\infty < \sigma_0 < \infty,\ \bar{\sigma} = \infty$). For Dirichlet's series of a given type, however, its breadth is subject to a certain limitation.

THEOREM 9. *We have* $\bar{\sigma} - \sigma_0 \leqq \overline{\lim} \dfrac{\log n}{\lambda_n}$.

We shall prove this theorem on the assumption that $\sigma_0 > 0$; its truth is obviously independent of this restriction. Given δ, we can choose n_0 so that

$$|A(n)| < e^{(\sigma_0+\delta)\lambda_n} \qquad (n \geqq n_0),$$

and accordingly

$$|a_n| = |A(n) - A(n-1)| < 2e^{(\sigma_0+\delta)\lambda_n} < e^{(\sigma_0+2\delta)\lambda_n}.*$$

Hence $$\bar{A}(n) = \sum_1^n |a_\nu| < \bar{A}(n_0) + n e^{(\sigma_0+2\delta)\lambda_n} < n e^{(\sigma_0+3\delta)\lambda_n}$$

if $n \geqq n_1$ and n_1 is sufficiently large in comparison with n_0. Thus

$$\frac{\log \bar{A}(n)}{\lambda_n} < \frac{\log n}{\lambda_n} + \sigma_0 + 3\delta \qquad (n \geqq n_1),$$

from which the theorem follows immediately.

If $\log n = o(\lambda_n)$, the lines of convergence and absolute convergence coincide: in particular this is the case if $\lambda_n = n$. In this case our theorems become, on effecting the transformation $e^{-s} = x$, classical theorems in the theory of power series. Thus Theorems 1 and 3 establish the existence of the circle of convergence, and 7 gives a slightly modified form of Cauchy's formula for the radius of convergence. Theorems 2, 4, 5, and 6 also become familiar results. If $\lambda_n = \log n$, the maximum possible distance between the lines of convergence is 1. This is of course an obvious consequence of the fact that $\sum n^{-1-\delta}$ is convergent for all positive values of δ.

It is not difficult to construct examples to show that every logically possible disposition of the lines of convergence and absolute convergence, consistent with Theorem 9, may actually occur. We content ourselves with mentioning the series

$$\sum \frac{(-1)^n}{\sqrt{n}} (\log n)^{-s},$$

which is convergent for all values of s, but never absolutely convergent.

8. It will be well at this point to call attention to the essential difference which distinguishes the general theory of Dirichlet's series from the simpler theory of power series, and lies at the root of the particular difficulties of the former. The region of convergence of a power series is determined in the simplest possible manner by the disposition of the singular points of the function which it represents: the circle of convergence extends up to the nearest singular point. As we shall see, no such simple relation holds in the general case; a Dirichlet's series convergent in a portion of the plane only may represent a function regular all over the plane, or in a wider region of

* If $e^{\delta\lambda_n} > 2$ for $n \geqq n_0$, as we can obviously suppose.

it. The result is (to put it roughly) that many of the peculiar difficulties which attend the study of power series *on* the circle of convergence are extended, in the case of Dirichlet's series, to wide regions of the plane or even to the whole of it.

There is however one important case in which the line of convergence necessarily contains at least one singularity.

THEOREM 10. *If all the coefficients of the series are positive or zero, then the real point of the line of convergence is a singular point of the function represented by the series**.

We may suppose that $\sigma_0 = \bar{\sigma} = 0$. Then, if $s = 0$ is a regular point, the Taylor's series for $f(s)$, at the point $s = 1$, has a radius of convergence greater than 1. Hence we can find a negative value of s for which

$$f(s) = \sum_{\nu=0}^{\infty} \frac{(s-1)^\nu}{\nu!} f^{(\nu)}(1) = \sum_{\nu=0}^{\infty} \frac{(1-s)^\nu}{\nu!} \sum_{n=1}^{\infty} a_n \lambda_n^\nu e^{-\lambda_n}.$$

But every term in this repeated series is positive. Hence† the order of summation may be inverted, and we obtain

$$f(s) = \sum_{n=1}^{\infty} a_n e^{-\lambda_n} \sum_{\nu=0}^{\infty} \frac{(1-s)^\nu \lambda_n^\nu}{\nu!} = \sum_{n=1}^{\infty} a_n e^{-\lambda_n s}.$$

Thus the series is convergent for some negative values of s, which contradicts our hypotheses.

In the general case all conceivable hypotheses may actually be realised. Thus the series

$$1^{-s} - 2^{-s} + 3^{-s} - \dots,$$

which converges for $\sigma > 0$, represents the function

$$(1 - 2^{1-s})\, \zeta(s), \ddagger$$

which is regular all over the plane. The series

$$\Sigma\, 2^{-2^n s}$$

has the imaginary axis as a line of essential singularities§.

* This theorem was proved first for power series by Vivanti, **1**, and Pringsheim, **1**. It was extended to the general case by Landau, **1**, and *H.*, p. 880. Further interesting generalisations have been made by Fekete, **1**, **2**.

† Bromwich, *Infinite series*, p. 78.

‡ For the theory of the famous ζ-function of Riemann, we must refer to Landau's *Handbuch* and the *Cambridge Tract* by Messrs Bohr and Littlewood which, we hope, is to follow this.

§ Landau, **2**. General classes of such series have been defined by Knopp **4**. Schnee, **1**, **3**, and Knopp, **1**, **3**, **5**, have also given a number of interesting theorems relating to the behaviour of $f(s)$ as s approaches a singular point on the line of convergence, the coefficients of the series being supposed to obey certain asymptotic laws. These theorems constitute a generalisation of the work of Appell, Cesàro, Lasker, Pringsheim and others on power series.

9. **Representation of a Dirichlet's series as a definite integral.** We may mention here the following theorem, which is interesting in itself and useful in the study of particular series. We shall not use it in this tract, and therefore do not include a proof.

THEOREM 11. *Let* $\mu_n = \log \lambda_n$. *Then*

$$\Sigma a_n e^{-\mu_n s} = \frac{1}{\Gamma(s)} \int_0^\infty x^{s-1} (\Sigma a_n e^{-\lambda_n x})\, dx$$

if $\sigma > 0$ *and the series on the left-hand side is convergent**.

We have, for example,

$$\zeta(s) = \Sigma n^{-s} = \frac{1}{\Gamma(s)} \int_0^\infty \frac{x^{s-1}}{e^x - 1}\, dx, \quad (1 - 2^{1-s})\, \zeta(s) = \Sigma (-1)^{n-1} n^{-s} = \frac{1}{\Gamma(s)} \int_0^\infty \frac{x^{s-1}}{e^x + 1}\, dx.$$

Here $\sigma > 1$ in the first formula, and $\sigma > 0$ in the second; and $\zeta(s)$ is the Riemann ζ-function.

III

THE FORMULA FOR THE SUM OF THE COEFFICIENTS OF A DIRICHLET'S SERIES: THE ORDER OF THE FUNCTION REPRESENTED BY THE SERIES

1. We shall now prove a theorem which is of fundamental importance for the later developments of the theory.

THEOREM 12†. *Suppose* $\lambda_1 \geqq 0$ *and the series convergent or finitely oscillating for* $s = \beta$. *Then*

$$\sum_1^n a_\nu e^{-\lambda_\nu s} = o\,|t|$$

uniformly for $\sigma \geqq \beta + \epsilon > \beta$ *and all values of* n; *that is to say, given any positive numbers* δ, ϵ, *we can find* t_0 *so that*

$$\left| \frac{1}{t} \sum_1^n a_\nu e^{-\lambda_\nu s} \right| < \delta$$

for $\sigma \geqq \beta + \epsilon$, $|t| \geqq t_0$, *and all values of* n. *In particular we have, for* $n = \infty$,

$$f(s) = o\,|t|$$

uniformly for $\sigma \geqq \beta + \epsilon$.

* See Cahen, **1**; Perron, **1**; Hardy, **5**; the last two authors give rigorous proofs.

† Landau, *H.*, p. 821.

We may take $\beta = 0$ without loss of generality. Then

$$|a_\nu| < K, \quad |A(\mu, \nu)| < K$$

for all values of μ and ν. Also, if $1 < N < n$, we have

$$\sum_1^n a_\nu e^{-\lambda_\nu s} = \sum_1^{N-1} a_\nu e^{-\lambda_\nu s} + \sum_N^{n-1} A(N, \nu)\, \Delta e^{-\lambda_\nu s} + A(N, n)\, e^{-\lambda_n s}$$
$$= S_1 + S_2 + S_3,$$

say; and since $|e^{-\lambda_n s}| < 1$ if $\sigma \geqq \epsilon$, we have

$$|S_1| < KN, \quad |S_3| < K, \quad S_1 + S_3 = O(N).$$

We have moreover, by Lemma 2 of II, § 1,

$$|S_2| < K\frac{|s|}{\sigma}\sum_N^{n-1} \Delta e^{-\lambda_\nu \sigma} < K\sqrt{\left(1 + \frac{t^2}{\epsilon^2}\right)}\, e^{-\lambda_N \epsilon},$$

$$\sum_1^n a_\nu e^{-\lambda_\nu s} = O(N) + O(te^{-\lambda_N \epsilon})$$

if $1 < N < n$. On the other hand it is evident that

$$\sum_1^n a_\nu e^{-\lambda_\nu s} = O(N)$$

if $N \geqq n$. If now we suppose that N is a function of $|t|$ which tends to infinity more slowly than $|t|$, we see that in any case

$$\sum_1^n a_\nu e^{-\lambda_\nu s} = o|t|.$$

2. We now apply Theorem 12 to prove an important theorem first rigorously and generally established by Perron*.

THEOREM 13. *If the series is convergent for* $s = \beta + i\gamma$, *and*

$$c > 0, \quad c > \beta, \quad \lambda_n < \omega < \lambda_{n+1},$$

then
$$\frac{1}{2\pi i}\int_{c-i\infty}^{c+i\infty} f(s)\, e^{\omega s}\frac{ds}{s} = \sum_1^n a_\nu,$$

the path of integration being the line $\sigma = c$. *At a point of discontinuity* $\omega = \lambda_n$, *the integral has a value half-way between its limits on either side, but in this case the integral must be regarded as being defined by its principal value*†.

* Perron, **1**. See also Cahen, **1**; Hadamard, **1** (where a rigorous proof is given for series which possess a half-plane of absolute convergence); Landau, *H.*, pp. 820 *et seq.*

† The *principal value* is the limit, if it exists, of

$$\frac{1}{2\pi i}\int_{c-iT}^{c+iT} f(s)\, e^{\omega s}\frac{ds}{s},$$

which may exist when the integral, as ordinarily defined, does not.

This theorem depends upon the following lemma.

LEMMA 3. *If x is real, we have*

$$\frac{1}{2\pi i}\int_{c-i\infty}^{c+i\infty} e^{xs}\frac{ds}{s} \begin{array}{ll} =1 & (x>0), \\ =\tfrac{1}{2} & (x=0), \\ =0 & (x<0), \end{array}$$

it being understood that in the second case the principal value of the integral is taken.

We may leave the verification of this result as an exercise to the reader*.

Let $\lambda_n < \omega < \lambda_{n+1}$ and

$$g(s) = e^{\omega s}\left\{f(s) - \sum_1^n a_\nu e^{-\lambda_\nu s}\right\} = \sum_{n+1}^\infty a_\nu e^{-(\lambda_\nu - \omega)s}$$

$$= \sum_1^\infty b_\nu e^{-\mu_\nu s},$$

where $b_\nu = a_{n+\nu}$, $\mu_\nu = \lambda_{n+\nu} - \omega$, so that $\mu_1 > 0$. It is clear from the lemma that what we have to show is that

$$\int_{c-i\infty}^{c+i\infty} g(s)\frac{ds}{s} = 0 \quad \text{.......................(1)}.$$

Applying Cauchy's theorem to the rectangle whose vertices are

$$c - iT_1,\ c + iT_2,\ \gamma + iT_2,\ \gamma - iT_1,$$

where T_1 and T_2 are positive, and $\gamma > c$, we obtain

$$\int_{c-iT_1}^{c+iT_2} g(s)\frac{ds}{s} = \int_{c-iT_1}^{\gamma-iT_1} g(s)\frac{ds}{s} - \int_{c+iT_2}^{\gamma+iT_2} g(s)\frac{ds}{s} + \int_{\gamma-iT_1}^{\gamma+iT_2} g(s)\frac{ds}{s}.$$

Keeping T_1 and T_2 fixed, we make γ tend to infinity. By Theorem 2, the upper limit of $|g(s)|$ in the last integral remains, throughout this process, less than a number independent of γ. Hence the last integral tends to zero, and

$$\int_{c-iT_1}^{c+iT_2} g(s)\frac{ds}{s} = \int_{c-iT_1}^{\infty-iT_1} g(s)\frac{ds}{s} - \int_{c+iT_2}^{\infty+iT_2} g(s)\frac{ds}{s} \text{.........(2)},$$

if the two integrals on the right-hand side are convergent. Now, if we write

$$g(s) = e^{-\mu_1 s} h(s),$$

* The easiest method of verification is by means of Cauchy's Theorem. Full details will be found in Landau, *H.*, pp. 342 *et seq.*

we can, by Theorem 12, choose T so that $|h(s)| < \epsilon T_2$ for $s = \sigma + iT_2$, $\sigma \geqq c$, $T_2 > T$. Hence the second integral on the right-hand side of (2) is convergent, and

$$\left| \int_{c+iT_2}^{\infty+iT_2} g(s) \frac{ds}{s} \right| < \frac{\epsilon T_2}{\sqrt{(c^2 + T_2^2)}} \int_c^\infty e^{-\mu_1 \xi} \, d\xi < \frac{\epsilon}{\mu_1}.$$

Thus the integral in question tends to zero as $T_2 \to \infty$. Similarly for the integral involving T_1. Hence (1) is established and the theorem is proved, except when ω is equal to one of the λ's. The reader will have no difficulty in supplying the modifications necessary in this case.

3. The order of $f(s)$ for $s = \beta$ and for $s \geqq \beta$. Theorem 12 suggests the introduction of an idea which will be prominent in the rest of this tract.

Suppose that $f(s)$ is a function of s regular for $\sigma > \gamma$. If $\beta > \gamma$, and ξ is any real number, it may or may not be true that

$$f(\sigma + ti) = O(|t|^\xi) \quad \ldots\ldots\ldots\ldots\ldots(1),$$

when $\sigma = \beta$ and $|t| \to \infty$. If this equation is true for a particular value of ξ, it is true for any greater value. It follows, by a classical argument, that there are three possibilities. The equation (1) may be true for all values of ξ, or for some but not all, or for none. In the second case there is a number μ such that (1) is true for $\xi > \mu$ and untrue for $\xi < \mu$. In the first case we may agree to write conventionally $\mu = -\infty$, and in the third case $\mu = \infty$. We thus obtain a function $\mu(\sigma)$ defined for $\sigma > \gamma$; and we call $\mu(\beta)$ *the order of $f(s)$ for $\sigma = \beta$*. When it is not true that $\mu(\beta) = \infty$, we say that $f(s)$ is *of finite order for $\sigma = \beta$*.

Again, the equation (1) may or may not hold uniformly for $\sigma \geqq \beta$. If we consider it from this point of view, and apply exactly the same arguments as before, we are led to define a function $\nu(\beta)$ which we call *the order of $f(s)$ for $\sigma \geqq \beta$*. Evidently $\nu \geqq \mu$. When it is not true that $\nu(\beta) = \infty$, we say that $f(s)$ is *of finite order for $\sigma \geqq \beta$*. And if $f(s)$ is of finite order for $\sigma \geqq \beta + \epsilon$, for every positive ϵ, but not necessarily for $\sigma \geqq \beta$, we shall say that it is *of finite order for $\sigma > \beta$*. Finally, the equation (1), without holding uniformly for $\sigma \geqq \beta_1$, may hold uniformly for $\beta_1 \leqq \sigma \leqq \beta_2$. We are thus led to define the order of $f(s)$ for $\beta_1 \leqq \sigma \leqq \beta_2$. The reader will find no difficulty in framing a formal definition, or in giving precise interpretations of the phrases '*$f(s)$ is of finite order for $\beta_1 \leqq \sigma \leqq \beta_2$*', '*$f(s)$ is of finite order for $\beta_1 < \sigma < \beta_2$*'.

4. Lindelöf's Theorem. In order to establish the fundamental properties of the function $\mu(\sigma)$ associated with a function $f(s)$, defined initially by a Dirichlet's series, we shall require the following theorem, which is due to Lindelöf, and is one of a class of general theorems the first of which were discovered by Phragmén *.

THEOREM 14. *If* (i) $f(s)$ *is regular and of finite order for* $\beta_1 \leqq \sigma \leqq \beta_2$, (ii) $f(s) = O(|t|^{k_1})$ *for* $\sigma = \beta_1$, (iii) $f(s) = O(|t|^{k_2})$ *for* $\sigma = \beta_2$, *then*

$$f(s) = O(|t|^{k(\sigma)}),$$

uniformly for $\beta_1 \leqq \sigma \leqq \beta_2$, $k(x)$ *being the linear function of* x *which assumes the values* k_1, k_2 *for* $x = \beta_1$, β_2.

The special case in which $k_1 = k_2 = 0$ is of particular interest; we have then the result that *if* $f(s)$ *is of finite order for* $\beta_1 \leqq \sigma \leqq \beta_2$, *and bounded on the lines* $\sigma = \beta_1$ *and* $\sigma = \beta_2$, *then it is bounded in the whole strip between them.*

In proving this theorem we may evidently confine our attention to positive values of t.

First, suppose k_1 as well as k_2 to be zero, so that $k(x)$ is identically zero and $f(s) = O(1)$ for $\sigma = \beta_1$ and $\sigma = \beta_2$. Let M be the upper bound of the values of $|f|$ on these two lines and the segment (β_1, β_2) of the real axis. Also let

$$g(s) = e^{\epsilon si} f(s) \qquad (\epsilon > 0).$$

Then
$$|g(s)| = e^{-\epsilon t} |f(s)| \leqq |f(s)|,$$

so that $g(s) = O(1)$ for $\sigma = \beta_1$ and $\sigma = \beta_2$. Also, as f is of finite order, $g \to 0$ as $t \to \infty$, uniformly for $\beta_1 \leqq \sigma \leqq \beta_2$. Hence, when ϵ is given, we can determine t_0 so that $|g| < M$ for $\beta_1 \leqq \sigma \leqq \beta_2$, $t > t_0$. It follows that any point whose abscissa lies between β_1 and β_2 can be surrounded by a contour at each point of which $|g| < M$, the contour being a rectangle formed by the lines $\sigma = \beta_1$, $\sigma = \beta_2$, the real axis, and a parallel to it at a sufficiently great distance from the origin. Hence, by a well-known theorem, $|g| < M$ at the point itself, and so

$$|f(s)| < Me^{\epsilon t}.$$

This is true for all positive values of ϵ, and therefore $|f| \leqq M$. Thus the theorem is proved. It should be observed that, if we had

* Lindelöf, **1**. See also Phragmén, **1**; Phragmén and Lindelöf, **1**; Landau, *H.*, pp. 849 *et seq.*

used the factor $e^{\epsilon s^2}$ instead of $e^{\epsilon si}$, we could have proved a little more, viz. that if $|f(s)|$ is less than M for $\sigma = \beta_1$ and $\sigma = \beta_2$, it is less than M for $\beta_1 \leqq \sigma \leqq \beta_2$. We leave the formal proof of this as an exercise for the reader.

Next, suppose $k(x)$ not identically zero, and consider the function*

$$h(s) = (-si)^{k(s)} = e^{k(s)\log(-si)},$$

where the logarithm has its principal value. This function is regular in the region $\beta_1 \leqq \sigma \leqq \beta_2$, $t \geqq 1$,† and, within this region, may be expressed in the form

$$e^{\{k(\sigma)+cti\}\{\log t + O(1/t)\}},$$

where c is a real constant. Thus

$$|h(s)| = t^{k(\sigma)}\, e^{O(1)},$$

so that the ratio of $|h(s)|$ to $t^{k(\sigma)}$ remains throughout the region between fixed positive limits. Hence the function

$$F(s) = f(s)/h(s)$$

satisfies the conditions which we supposed before to be satisfied by $f(s)$. Thus

$$F(s) = O(1), \quad f(s) = O\{t^{k(\sigma)}\},$$

uniformly throughout the region; and the theorem is completely proved.

5. Properties of the function $\mu(\sigma)$ associated with a Dirichlet's series which has a domain of absolute convergence. We shall now apply Lindelöf's Theorem to establish the fundamental properties of the function $\mu(\sigma)$, when $f(s)$ is defined by a Dirichlet's series. In order to obtain simple and definite results, we shall limit ourselves to the case in which there is a domain of absolute convergence.

THEOREM 15. *Suppose that the series* $\sum a_n e^{-\lambda_n s}$ *is absolutely convergent for* $\sigma > \bar{\sigma}$, *and that the function* $f(s)$, *defined by the series when* $\sigma > \bar{\sigma}$, *is regular and of finite order for* $\sigma > \gamma$, *where* $\gamma < \bar{\sigma}$. *Then the function* $\mu(\sigma)$, *defined for* $\sigma > \gamma$, *has the following properties. Either it is always zero; or it is zero for* $\sigma \geqq \gamma_0$, *where* $\gamma < \gamma_0 \leqq \bar{\sigma}$, *while*

* This auxiliary function (introduced by Landau, 9) is a little simpler than that used by Lindelöf.

† We suppose $t \geqq 1$ instead of, as before, $t \geqq 0$, to avoid the singularity of $h(s)$ for $s = 0$ even when $\beta_1 \leqq 0 \leqq \beta_2$. In proving the theorem it is evidently only necessary to consider values of t greater than some fixed value.

for $\gamma < \sigma < \gamma_0$ *it is a positive, decreasing, convex*, and continuous function of* σ. *Further,* $\nu(\sigma)$ *is identical with* $\mu(\sigma)$.†

Suppose that $\mu = \mu_1$ for $\sigma = \beta_1 > \gamma$, and $\mu = \mu_2$ for $\sigma = \beta_2 > \beta_1$. Then

$$f(\beta_1 + ti) = O(|t|^{\mu_1+\epsilon_1}), \quad f(\beta_2 + ti) = O(|t|^{\mu_2+\epsilon_2}),$$

where ϵ_1, ϵ_2 are any positive numbers. Applying Lindelöf's Theorem we obtain at once

$$\mu \leq \frac{\beta_2 - \sigma}{\beta_2 - \beta_1}(\mu_1 + \epsilon_1) + \frac{\sigma - \beta_1}{\beta_2 - \beta_1}(\mu_2 + \epsilon_2),$$

or, as ϵ_1 and ϵ_2 are arbitrarily small,

$$\mu \leq \frac{(\beta_2 - \sigma)\mu_1 + (\sigma - \beta_1)\mu_2}{\beta_2 - \beta_1} \quad \text{.....................(1)},$$

for $\beta_1 \leq \sigma \leq \beta_2$. This relation expresses the fundamental property of the function μ.

A similar argument shows that, if $\mu = -\infty$ for any σ, the same must be true for every σ. We shall see in a moment that this possibility may, in the present case, be ignored.

It is clear that $\mu \leqq 0$ for $\sigma > \bar{\sigma}$; for if $\beta > \bar{\sigma}$ then

$$|f(s)| < \Sigma |a_n| e^{-\lambda_n \beta}$$

for $\sigma \geqq \beta$. But it is easy to see also that $\mu \geqq 0$ for sufficiently large values of σ. For, if a_m is the first coefficient in the series which does not vanish, we may write $f(s)$ in the form

$$a_m e^{-\lambda_m s} + e^{-\lambda_m s} \sum_{n>m} a_n e^{-(\lambda_n - \lambda_m)s}.$$

The series here written is absolutely and uniformly convergent for $\sigma > \bar{\sigma}$, and so tends uniformly to zero as $\sigma \to \infty$. Hence we can so choose ω that

$$f(s) = a_m e^{-\lambda_m s}(1 + \rho),$$

* We say that $f(x)$ is *convex* if it satisfies the inequality

$$f\{\theta x + (1-\theta) y\} \leqq \theta f(x) + (1-\theta) f(y)$$

for $0 \leqq \theta \leqq 1$. The theory of such functions has been investigated systematically by Jensen, **3**. If we put $\theta = \frac{1}{2}$ we obtain the inequality

$$2f\{\tfrac{1}{2}(x+y)\} \leqq f(x) + f(y);$$

and Jensen has shown that, if $f(x)$ is continuous, the more general inequality can be deduced from this. A continuous function is certainly convex if

$$\lim_{h \to 0} \frac{f(x+h) - 2f(x) + f(x-h)}{h^2}$$

exists for all values of x and is never negative. See Harnack, **1**, and Hölder, **1**.

† The results comprised in Theorem 15 are in the main due to Bohr, **5**.

where $|\rho| < \frac{1}{2}$ for $\sigma > \omega$. Thus $|f|$ has a positive lower bound when σ has a fixed value greater than ω and $|t| \to \infty$; and so $\mu \geqq 0$, and therefore $\mu = 0$, for $\sigma > \omega$.

We can now show that μ can never be negative. For if μ were ever negative we could suppose, in (1), that $\mu_1 < 0$, while β_2 and σ are both greater than ω, so that $\mu_2 = 0$, $\mu = 0$. This obviously involves a contradiction. We thus see that $\mu = 0$ for $\sigma > \bar{\sigma}$.

Again, if in (1) we suppose $\mu_1 > 0$, and $\beta_2 > \bar{\sigma}$, so that $\mu_2 = 0$, we see that $\mu < \mu_1$ if $\sigma > \beta_1$. Thus μ, in so far as it is not zero, is a decreasing function of σ, in the stricter sense which forbids equality of values.

The only property of $\mu(\sigma)$ which remains to be established is its continuity. If β_1 is a particular value of σ, the numbers

$$\mu(\beta_1 - 0) = \mu_1', \quad \mu(\beta_1) = \mu_1, \quad \mu(\beta_1 + 0) = \mu_1''$$

all exist (since μ is monotonic) and $\mu_1' \geqq \mu_1 \geqq \mu_1''$. But since μ is convex, we have the inequalities

$$\mu(\beta_1 - \delta) \leqq \mu(\beta_1 - 2\delta) - \mu(\beta_1 - \delta) + \mu(\beta_1),$$
$$\mu(\beta_1) - \mu(\beta_1 + \delta) \leqq \mu(\beta_1 - \delta) - \mu(\beta_1),$$

where δ is positive. Making $\delta \to 0$, we obtain from the first $\mu_1' \leqq \mu_1$, and then from the second $\mu_1 \leqq \mu_1''$. It follows that $\mu_1' = \mu_1 = \mu_1''$, so that μ is continuous.

Finally, in order to establish the equivalence of the functions μ and ν, it is only necessary to apply Theorem 14 to a strip whose right-hand edge lies in the domain of absolute convergence, and to take account of the uniformity asserted by the theorem. This we may leave to the reader.

We may remark that it follows from Theorem 12 that $\mu \leqq 1$ for $\sigma > \sigma_0$, where σ_0 is the abscissa of convergence.

6. The actual determination of the function $\mu(\sigma)$ associated with a given Dirichlet's series is in general a problem of extreme difficulty Consider for example the series

$$1^{-s} - 2^{-s} + 3^{-s} - \ldots = (1 - 2^{1-s})\zeta(s).$$

The series is convergent for $\sigma > 0$, and absolutely convergent for $\sigma > 1$; the function is regular all over the plane. Obviously $\mu(\sigma) = 0$ for $\sigma \geqq 1$, while it may easily be shown, by means of Riemann's functional equation for the ζ-function, that $\mu(\sigma) = \frac{1}{2} - \sigma$ for $\sigma \leqq 0$. All that is known about $\mu(\sigma)$ for $0 \leqq \sigma \leqq 1$ is that its graph does not rise above the line joining the points $(0, \frac{1}{2})$ and $(1, 0)$*. It has however been proved by Littlewood† that, if it be true that all the complex roots of $\zeta(s)$ have the real part $\frac{1}{2}$, then

$$\mu(\sigma) = \frac{1}{2} - \sigma \quad (\sigma < \tfrac{1}{2}), \quad \mu(\sigma) = 0 \quad (\sigma \geqq \tfrac{1}{2}).$$

* Lindelöf, **1**; Landau, *H.*, p. 868. † Littlewood, **2**.

7. Let us consider now a Dirichlet's series with distinct lines of convergence and absolute convergence. To fix our ideas let us suppose $\sigma_0=0$, $\bar{\sigma}=1$, and the function regular and of finite order for some negative values of σ. Then $\mu=0$ for $\sigma=1$, and $\mu\leqq 1$ for $\sigma=0$. It follows at once, from the convexity of μ, that $\mu\leqq 1-\sigma$ for $0\leqq\sigma\leqq 1$. It will be proved later on* that $\mu>0$ for $\sigma<0$. Thus the final range of invariability of μ, which cannot begin later than $\sigma=1$, cannot begin earlier than $\sigma=0$. Bohr† has constructed examples which show that the two extreme cases thus indicated can actually occur. He has shown that it is possible to find two ordinary Dirichlet's series for each of which $\sigma_0=0$, $\bar{\sigma}=1$, while for one series $\mu=0$ for $\sigma>0$ and for the other $\mu=1-\sigma$ for $0<\sigma<1$. He has also shown‡ that it is possible for two ordinary Dirichlet's series to have the same μ-function but different regions of convergence, and so that it is futile to attempt to define the region of convergence of a Dirichlet's series in terms merely of the associated μ-function.

So far we have assumed the existence of a domain of absolute convergence. Some of our arguments remain valid in the general case, but it is no longer possible to obtain such simple and satisfactory results. We shall content ourselves, therefore, with mentioning one further interesting result of Bohr. Suppose that the indices λ_n are *linearly independent* §, that is to say that there are no relations of the type

$$k_1\lambda_1+k_2\lambda_2+\ldots+k_n\lambda_n=0,$$

where the k's are integers, not all zero, holding between them. Then Bohr|| has shown that, if $f(s)$ is regular and *bounded* for $\sigma\geqq\beta$, the series is absolutely convergent for $\sigma\geqq\beta$. Thus if there is no region of absolute convergence, the function cannot be bounded in any half-plane. He has also shown by an example that this conclusion is no longer necessarily correct when the restriction of the linear independence of the λ's is removed.

IV

THE SUMMATION OF SERIES BY TYPICAL MEANS

1. So far we have considered only *convergent* Dirichlet's series. We have seen that such a series defines an analytical function which may or may not exist outside of the domain of convergence of the series.

* See VII, § 10, Theorem 50. † Bohr, **5**, pp. 30 *et seq.*

‡ *l.c. supra* p. 36.

§ This is in a sense the *general* case. The condition is satisfied, for example, when $\lambda_n=\log p_n$, where p_n is the n-th prime, but not when $\lambda_n=n$ or $\lambda_n=\log n$. The *result* of course still holds when $\lambda_n=n$.

|| Bohr, **7**.

In the modern developments of the theory of power series a great part has been played by a variety of methods of summation of oscillating series, which we associate with the names of Frobenius, Hölder, Cesàro, Borel, Lindelöf, Mittag-Leffler, and Le Roy*. Of these definitions the simplest and the most natural is that which defines the sum of an oscillating series as the limit of the arithmetic mean of its first n partial sums. This definition was generalised in two different ways by Hölder and by Cesàro, who thus arrived at two systems of definitions the complete equivalence of which has been established only recently by Knopp, Schnee, Ford, and Schur†.

The range of application of Cesàro's methods is limited in a way which forbids their application to the problem of the analytical continuation of the function represented by a Taylor's series. A power series, outside its circle of convergence, diverges too crudely for the application of such methods: more powerful, though less delicate, methods, such as Borel's ‡, are required. But Cesàro's methods have proved of the highest value in the study of power series *on* the circle of convergence and the closely connected problems of the theory of Fourier's series§. And it is natural to suppose that in the theory of Dirichlet's series, where we are dealing with series whose convergence or divergence is of a much more delicate character than is, in general, that of a power series, they will find a wider field of application.

The first such applications were made independently by Bohr and Riesz‖, who showed that the arithmetic means formed in Cesàro's manner from an ordinary Dirichlet's series may have domains of convergence more extensive than that of the series itself¶. But it appeared from the investigations of Riesz that these arithmetic means

* For a general account of some of these methods and the relations between them, see Borel, *Leçons sur les séries divergentes*, Ch. 3; Bromwich, *Infinite series*, Ch. 11; Hardy and Chapman, **1**; Hardy and Littlewood, **1**, **2**.

† Knopp, **1**; Schnee, **2**; Ford, **1**; Schur, **1**. See also Bromwich, **1**; Faber, **1**; Landau, **13**; Knopp, **7**.

‡ Cf. Hardy and Littlewood, **1**.

§ We need only mention Fejér's well-known theorem (Fejér, **1**) and its generalisations by Lebesgue, **1**; Riesz, **3**; Chapman, **1**, **2**; Young, **1**, **2**; Hardy and Littlewood, **3**; and Hardy, **9**. We should say here that when referring to Marçel Riesz we write simply 'Riesz'.

‖ Bohr, **1**, **2**, **5**, **6**; Riesz, **1**, **2**, **3**, **4**.

¶ Thus, as was shown in a very simple manner by Bohr, **5**, the series

$$\Sigma(-1)^{n-1}n^{-s}$$

is summable by Cesàro's k-th mean if $\sigma > -k$. In so far as real values of s are concerned, this had already been proved by Cesàro, in a somewhat less elementary way (see Bromwich, *Infinite series*, p. 317). The general result follows from this and Theorem 29 below.

are not so well adapted to the study of the series as certain other means formed in a somewhat different manner. These 'logarithmic means' *, as well as the arithmetic means, have generalisations especially adapted to the study of the general series $\Sigma a_n e^{-\lambda_n s}$. We shall begin by giving the formal definitions of these means †; we shall then indicate shortly how they form a natural generalisation of Cesàro's.

2. Definitions. We suppose $\lambda_1 \geqq 0$, and we write

$$e^{\lambda_n} = l_n, \quad a_n e^{-\lambda_n s} = a_n l_n^{-s} = c_n,$$

$$C_\lambda(\tau) = \sum_{\lambda_n < \tau} c_n, \quad C_l(t) = \sum_{l_n < t} c_n.$$

Thus $$C_\lambda(\tau) = c_1 + c_2 + \dots + c_n \qquad (\lambda_n < \tau \leqq \lambda_{n+1}).$$

If $\lambda_n = n$, $C_\lambda(\tau)$ is identical with the function $C(\tau)$ defined in I, § 2, except when τ is an integer n, when the two functions differ by c_n. ‡

Further, we shall write

$$C_\lambda^\kappa(\omega) = \sum_{\lambda_n < \omega} (\omega - \lambda_n)^\kappa c_n = \kappa \int_0^\omega C_\lambda(\tau)(\omega - \tau)^{\kappa - 1} d\tau,$$

where κ is any positive number, integral or not§. We leave it to the reader to verify the equivalence of the two expressions of $C_\lambda^\kappa(\omega)$. In a precisely similar way we define $C_l^\kappa(w)$: thus

$$C_l^\kappa(w) = \sum_{l_n < w} (w - l_n)^\kappa c_n = \kappa \int_1^w C_l(t)(w - t)^{\kappa - 1} dt. \|$$

We shall call the functions

$$C_\lambda^\kappa(\omega)/\omega^\kappa, \quad C_l^\kappa(w)/w^\kappa,$$

introduced by Riesz, the *typical means* (*moyennes typiques*¶) *of order* κ,

* Riesz, **2**.

† Riesz, **3**. The definitions of this note, which are those which we adopt as final, differ from those of the earlier note.

‡ If $\lambda(x)$ is an increasing function of x, which assumes the value λ_n for $x = n$, then $C_\lambda(\lambda)$, regarded as a function of x, is, except for integral values of x, the same function as $C(x)$.

§ The theory of non-integral orders of summation by means of Cesàro's type has been developed by Knopp, **1**, **2**, and by Chapman, **1**. These writers consider also *negative* orders of summation (greater than -1): but we shall not be concerned with such negative orders here. It should be observed that $C_\lambda^\kappa(\omega) = 0$ if $\omega \leqslant \lambda_1$.

‖ We have $C_l(t) = 0$ if $t \leqq l_1$; and $l_1 = e^{\lambda_1} \geqslant 1$. Thus we may take 1 as the lower limit. And $C_l^\kappa(w) = 0$ if $w \leqslant l_1$.

¶ Riesz, **3**. The type is the type of the associated Dirichlet's series (I, § 1).

of the first and the second kind, associated with the series $\Sigma a_n e^{-\lambda_n s}$. It should be observed that, so long as we are thinking merely of the problem of summing a numerical series Σc_n, the word 'typical' is devoid of significance; it only acquires a significance when we are summing a Dirichlet's series of a special type by means specially defined with reference to that type. We shall frequently omit the suffixes λ, l, when no ambiguity can arise from so doing. The reader must however bear in mind the distinction already referred to between the function $C(\tau)$ thus defined and the function $C(\tau)$ of Section I.

If

$$\omega^{-\kappa} C_\lambda^\kappa(\omega) \rightarrow C$$

as $\omega \rightarrow \infty$, *we shall say that the series* Σc_n *is summable* (λ, κ) *to sum* C*. If the typical mean oscillates finitely as $\omega \rightarrow \infty$, we shall say that the series is *finite* (λ, κ). Similarly we define summability and finitude (l, κ).

We add a few remarks† to show the genesis of these definitions. If $\lambda_n = n$, the typical mean of the first kind is

$$\omega^{-\kappa} \sum_{n<\omega} (\omega - n)^\kappa c_n = \kappa \omega^{-\kappa} \int_0^\omega C(\tau)(\omega - \tau)^{\kappa-1} d\tau.$$

If in particular $\kappa = 1$, and ω is an integer, we obtain

$$\frac{1}{\omega} \sum_1^{\omega-1} (\omega - n) c_n = \frac{C_1 + C_2 + \ldots + C_{\omega-1}}{\omega}$$

(where $C_n = c_1 + c_2 + \ldots + c_n$), which is practically Cesàro's first mean. If κ is an integer greater than unity, we have

$$\kappa \omega^{-\kappa} \int_0^\omega C(\tau)(\omega - \tau)^{\kappa-1} d\tau = \kappa! \, \omega^{-\kappa} \left(\int_0 d\tau\right)^\kappa C(\tau).‡$$

Now Cesàro's κ-th mean is

$$\kappa! \, n^{-\kappa} C_n^\kappa,$$

where C_n^κ is the κ-th repeated sum formed from the numbers C_n, and it is plain that, as soon as we replace C_n by a function $C(\tau)$ of a continuous variable τ, we are naturally led to the definitions adopted here. And it is then also natural to abandon the restriction that κ is an integer. The integrals to which we are thus led are of course of the type employed by Liouville and Riemann in their theories of non-integral orders of differentiation and integration §.

* This is substantially the notation introduced by Hardy, **4**. Hardy writes '*summable* (R, λ, κ)'—*i.e.* summable by Riesz's means of type λ and order κ.

† Compare Hardy, **6**.

‡ See, *e.g.*, Jordan, *Cours d'analyse*, Vol. 3, p. 59.

§ Liouville, **1**, **2**; Riemann, **1**. See also Borel, *Leçons sur les séries à termes positifs*, pp. 74 *et seq.*

It has in fact been shown by Riesz* that these definitions are completely equivalent to Cesàro's, and to the generalisations of Cesàro's considered by Knopp and Chapman†. It is this which justifies our calling the typical means of the first kind, when $\lambda_n = n$, *arithmetic* means. In this case the means of the second kind are not of any interest‡.

So much for the case when $\lambda_n = n$. If in the general definition we put $\kappa = 1$, $\omega = \lambda_n$, we obtain

$$(\mu_1 C_1 + \mu_2 C_2 + \ldots + \mu_{n-1} C_{n-1})/\lambda_n,$$

where $\mu_\nu = \lambda_{\nu+1} - \lambda_\nu$. This is the natural generalisation of Cesàro's first mean which suggests itself when we try to attach varying weights to the successive partial sums C_ν.

When $\lambda_n = \log n$, the series $\Sigma a_n e^{-\lambda_n s}$ is an ordinary Dirichlet's series. The means of the first kind are then what Riesz has called *logarithmic means*§, and it is the means of the second kind that are arithmetic means. From the theoretical standpoint, the former are in general better adapted to the study of ordinary Dirichlet's series‖. On the other hand, arithmetic means are simpler in form and often easier to work with. Hence it is convenient and indeed necessary to take account of both kinds of means; and the same is true, of course, in the general theory.

3. Summable Integrals. It is easy to frame corresponding definitions for integrals. We suppose that $\lambda(x)$ is a positive and continuous function of x, which tends steadily to infinity with x, and that $\lambda(0) = 0$, and write

$$C_\lambda(x) = \int_{\lambda(u)<x} c(u)\,du = \int_0^{\bar\lambda(x)} c(u)\,du,$$

where $\bar\lambda$ is the function inverse to λ. Further we write

$$C_\lambda^\kappa(\omega) = \kappa \int_0^\omega C_\lambda(x)(\omega - x)^{\kappa-1}\,dx.$$

Then *if*
$$\omega^{-\kappa} C_\lambda^\kappa(\omega) \to C,$$

as $\omega \to \infty$, *we shall say that the integral*

$$\int_0^\infty c(x)\,dx$$

is summable (λ, κ) *to sum* C.

This definition may be applied to the theory of integrals of the type

$$\int_0^\infty a(x)\,e^{-s\lambda(x)}\,dx$$

* Riesz, **4**.

† Knopp, **1**, **2**; Chapman, **1**.

‡ See § 4 (3).

§ Riesz, **2**.

‖ It may also be observed that the form of the arithmetic means which we have adopted is better adapted for this purpose than that of Cesàro. Thus, for example, Schnee (**7**, pp. 393 *et seq.*), working with Cesàro's means, was able to avoid an unnecessary restriction only by using the result of Riesz cited above as to the equivalence of Cesàro's means and our 'arithmetic' means. See p. 56, footnote (*).

in the same way that the definitions of § 2 may be applied to the theory of Dirichlet's series*.

Most of the theorems which we shall prove for series have their analogues for integrals. When this is so, the proofs are in general easier for integrals than for series, and we have not thought it worth while to give the details of any of them. We have not even stated the theorems themselves explicitly, except one theorem (V, § 3) of which we shall make several applications.

4. In this paragraph we shall state, without detailed proofs, a number of special results which may be regarded as exercises for the reader. Some of them are in reality special cases of general theorems which we shall give later on. They are inserted here, partly because they are interesting in themselves, and partly in order to familiarise the reader with our notation and to give him a general idea of the range of our definitions.

(1) If Σa_n is convergent and has the sum A, then

$$\Sigma a_n e^{-\lambda_n s} = A + \int_0^\infty s e^{-su} \{A_\lambda(u) - A\}\, du$$

for all values of s for which the series is convergent. If in addition $\sigma > 0$, then

$$\Sigma a_n e^{-\lambda_n s} = \int_0^\infty s e^{-su} A_\lambda(u)\, du.$$

These formulae are simple examples of a mode of representation of Dirichlet's series by integrals which we shall often have occasion to use.

(2) Every convergent series is summable; more generally, the limits of oscillation of the typical means associated with a series are at most as wide as those of the series itself. When the series is given, it is possible to choose a sequence (λ_n) in such a way that the series shall be summable $(\lambda, 1)$ and have as its sum any number which does not lie outside its limits of oscillation.

(3) When $\lambda_n = e^n$, a series is summable (λ, κ) if and only if it is convergent. It follows from Theorem 21 below that this is true whenever $\lambda_{n+1}/\lambda_n \geqq k > 1$ for all values of n.

(4) The series $$1^{-s} - 2^{-s} + 3^{-s} - \dots$$ is summable (n, κ) if and only if $\kappa > -\sigma$. In other words, it is summable by Cesàro's κ-th mean if and only if $\kappa > -\sigma$. For references to proofs of this proposition, when κ is an integer, see the footnote (¶) to p. 20. The general result is distinctly more difficult to prove; a direct proof has been given by Chapman, and another indicated by Hardy and Chapman†.

* The integral may obviously be reduced, by the substitution $\lambda(x) = \xi$, to one in which the exponential factor has the simpler form $e^{-s\xi}$. There is no such fundamental distinction of integrals into types as there is with series. But an integral may be summable (λ, κ) for one form of λ and not for another.

† Chapman, **1**; Hardy and Chapman, **1**.

Exactly the same result holds for the more general series* $\Sigma e^{ani} n^{-s}$ (provided a is not a multiple of 2π) and for the integral†

$$\int_0^\infty e^{axi} x^{-s}\, dx.$$

(5) The series $\qquad \Sigma e^{Ain^a} n^{-s} \qquad (0<a<1,\ A \neq 0)$

is summable (n, κ) if $(\kappa+1)a+\sigma>1$.‡

(6) The series $\qquad \Sigma n^{-1-ti} \qquad (t \neq 0)$

is not summable (n, κ) for any value of κ, but is summable $(\log n, \kappa)$ for any positive κ however small§. The series

$$\Sigma n^{-1} (\log n)^{-1-ti}$$

is not summable $(\log n, \kappa)$ for any value of κ, but is summable $(\log\log n, \kappa)$ for any positive κ however small‖; and so on generally.

(7) The series Σc_n is summable $(\log n, 1)$ to sum C if and only if

$$\left(C_1 + \frac{1}{2} C_2 + \ldots + \frac{1}{n} C_n\right) \Big/ \log n \rightarrow C.¶$$

(8) If Σa_n is summable (λ, κ) to sum A, and Σb_n is summable (λ, κ) to sum B, then $\Sigma(pa_n + qb_n)$ is summable (λ, κ) to sum $pA+qB$.

(9) If $\Sigma a_n e^{-\lambda_n s}$ is summable (λ, κ) to sum $f(s)$, then

$$a_{m+1} e^{-\lambda_{m+1} s} + a_{m+2} e^{-\lambda_{m+2} s} + \ldots$$

is summable (μ, κ), where $\mu_n = \lambda_{m+n}$, to sum

$$f(s) - a_1 e^{-\lambda_1 s} - \ldots - a_m e^{-\lambda_m s}.$$

(10) If $\Sigma a_n e^{-\lambda_n s}$ is summable (λ, κ) to sum $f(s)$, then $\Sigma a_n e^{-(\lambda_n - \lambda_1)s}$ is summable (μ, κ), where $\mu_n = \lambda_n - \lambda_1$, to sum $e^{\lambda_1 s} f(s)$.

The last two examples will be used later (VII, § 2). The first is a corollary of (8). The second follows from the identity

$$\sum_{\lambda_n - \lambda_1 < \omega} \left(1 - \frac{\lambda_n - \lambda_1}{\omega}\right)^\kappa a_n e^{-(\lambda_n - \lambda_1)s}$$

$$= \left(\frac{\omega + \lambda_1}{\omega}\right)^\kappa e^{\lambda_1 s} \sum_{\lambda_n < \omega + \lambda_1} \left(1 - \frac{\lambda_n}{\omega + \lambda_1}\right)^\kappa a_n e^{-\lambda_n s}.$$

* Chapman, 1. † Hardy, 6.

‡ Hardy, 6 (where A is taken to be 1). This paper contains a number of general theorems concerning the relations, from the point of view of convergence and summability, between the series $\Sigma f(n)$ and the integral $\int f(x)\, dx$.

§ Riesz, 1. See also Hardy, 4, and Theorems 19, 42, and 47.

‖ Hardy, 4. ¶ Riesz, 2.

(11) The summability of the series

$$c_1+c_2+c_3+\dots$$

does not, in general, involve that of the series

$$c_2+c_3+c_4+\dots;$$

nor is the converse proposition true. It may even happen that both series are summable and their sums do not differ by c_1. Both propositions are true, however, if the increase of λ_n is sufficiently regular, and in particular if λ_n is a logarithmico-exponential function*, *i.e.* a function defined by any finite combination of logarithms and exponentials.

V

GENERAL ARITHMETIC THEOREMS CONCERNING TYPICAL MEANS

1. The general theorems which we shall prove concerning the summation of series by typical means may be divided roughly into two classes. There are, in the first place, theorems the validity of which does not depend upon any hypothesis that the series considered are Dirichlet's series of any special type. Such theorems we may call 'arithmetic'. There are other theorems in which such a hypothesis is essential. Thus Theorem 23 of Section VI depends upon the fact that we are applying methods of summation of type λ to a series of the same type. Such a theorem we may describe as 'typical'†.

The theorems of this section are all 'arithmetic', those of the following sections mainly (though not entirely) 'typical'.

2. Five lemmas. We shall now give five lemmas which will be useful in the sequel.

LEMMA 4. *If ϕ is a positive function of x such that*

$$\int^{\infty} \phi\, dx$$

is divergent, and $f=o(\phi)$, then

$$\int^{x} f\, dx = o\left(\int^{x} \phi\, dx\right).$$

The proof may be left to the reader.

* Hardy, *Orders of infinity*, p. 17.

† We make this distinction merely for the sake of convenience in exposition and lay no stress upon it.

LEMMA 5. *Let $\phi(x)$, $\psi(x)$ be continuous functions of x such that*

$$\phi(x) \sim Ax^{\alpha}, \quad \psi(x) \sim Bx^{\beta} \qquad (\alpha \geqq 0,\ \beta \geqq 0)$$

as $x \to \infty$. Then

$$\chi(x) = \int_0^x \phi(t)\,\psi(x-t)\,dt \sim AB\frac{\Gamma(\alpha+1)\,\Gamma(\beta+1)}{\Gamma(\alpha+\beta+2)}\,x^{\alpha+\beta+1}.*$$

We can write

$$\phi(t) = At^{\alpha} + \phi_1, \quad \psi(t) = Bt^{\beta} + \psi_1,$$

where

$$\phi_1 = o(t^{\alpha}), \quad \psi_1 = o(t^{\beta}).$$

If we substitute these forms of ϕ and ψ in χ, we obtain a sum of four integrals, the first of which gives us

$$AB\int_0^x t^{\alpha}(x-t)^{\beta}\,dt = AB\,\frac{\Gamma(\alpha+1)\,\Gamma(\beta+1)}{\Gamma(\alpha+\beta+2)}\,x^{\alpha+\beta+1}.$$

It remains to prove that the other three integrals are of the form

$$o(x^{\alpha+\beta+1}).$$

Let us take, for example, the integral

$$A\int_0^x t^{\alpha}\,\psi_1(x-t)\,dt = A\int_0^x (x-t)^{\alpha}\,\psi_1(t)\,dt.$$

Given ϵ, we can choose ξ so that

$$|\psi_1(t)| < \epsilon t^{\beta} \qquad (\xi \leqq t \leqq x).$$

Between 0 and ξ, $|\psi_1(t)|$ is less than a number $M(\xi)$ which depends only on ξ. Hence our integral is in absolute value less than

$$|A|\left\{\epsilon\int_0^x (x-t)^{\alpha}\,t^{\beta}\,dt + \xi M(\xi)\,x^{\alpha}\right\},$$

and is therefore of the form $o(x^{\alpha+\beta+1})$. The other integrals may be proved similarly to be also of this form, and thus the proof of the lemma is completed.

LEMMA 6. *If $\kappa > 0$, $\mu > 0$, then*

$$C^{\kappa+\mu}(\omega) = \frac{\Gamma(\kappa+\mu+1)}{\Gamma(\kappa+1)\,\Gamma(\mu)}\int_0^{\omega} C^{\kappa}(u)\,(\omega-u)^{\mu-1}\,du.$$

And if $\kappa > 0$, $\mu < 1$, $\mu \leqq \kappa$, then

$$C^{\kappa-\mu}(\omega) = \frac{\Gamma(\kappa-\mu+1)}{\Gamma(\kappa+1)\,\Gamma(1-\mu)}\int_0^{\omega}\frac{dC^{\kappa}(u)}{du}\,(\omega-u)^{-\mu}\,du.$$

* Chapman, 1. The result is true for $\alpha > -1$, $\beta > -1$: we state it in the form in which we shall use it. By $\phi(x) \sim Ax^{\alpha}$ we mean that $\phi/x^{\alpha} \to A$ as $x \to \infty$.

To prove the first formula, we substitute for $C^\kappa(u)$ its expression as a definite integral*. We thus obtain

$$\int_0^\omega C^\kappa(u)(\omega-u)^{\mu-1}\,du = \kappa\int_0^\omega (\omega-u)^{\mu-1}\,du\int_0^u C(\tau)(u-\tau)^{\kappa-1}\,d\tau$$

$$= \kappa\int_0^\omega C(\tau)\,d\tau\int_\tau^\omega (u-\tau)^{\kappa-1}(\omega-u)^{\mu-1}\,du$$

$$= \frac{\Gamma(\kappa+1)\Gamma(\mu)}{\Gamma(\kappa+\mu)}\int_0^\omega C(\tau)(\omega-\tau)^{\kappa+\mu-1}\,d\tau,$$

which is the formula required. The legitimacy of the inversion of the order of integration follows at once from classical theorems †.

When μ is a positive integer, we have the simpler formula

$$C^{\kappa+\mu}(\omega) = (\kappa+1)(\kappa+2)\dots(\kappa+\mu)\left(\int_0^\omega d\tau\right)^\mu C^\kappa(\tau).$$

To prove the second formula of the lemma, we observe that

$$C^{\kappa-\mu}(\omega) = \frac{1}{\kappa-\mu+1}\frac{d}{d\omega}C^{\kappa-\mu+1}(\omega)$$

$$= \frac{\Gamma(\kappa-\mu+1)}{\Gamma(\kappa+1)\Gamma(1-\mu)}\frac{d}{d\omega}\int_0^\omega C^\kappa(u)(\omega-u)^{-\mu}\,du,$$

by the first formula. Integrating by parts, we find

$$\int_0^\omega C^\kappa(u)(\omega-u)^{-\mu}\,du = \frac{1}{1-\mu}\int_0^\omega \frac{dC^\kappa(u)}{du}(\omega-u)^{1-\mu}\,du.$$

Differentiating with respect to ω‡ we obtain the second formula.

LEMMA 7. *If c_n is real, $0 \leq \xi \leq \omega$, and $0 < \kappa \leq 1$, then*

$$|g(\xi,\omega)| = \left|\kappa\int_0^\xi C(\tau)(\omega-\tau)^{\kappa-1}\,d\tau\right| \leq \max_{0\leqq\tau\leqq\xi}|C^\kappa(\tau)|.$$

The truth of the lemma is evident when $\kappa = 1$, and we may therefore suppose $\kappa < 1$. Substituting in the integral which defines $g(\xi,\omega)$ the expression of $C(\tau)$ as an integral given by the second formula of Lemma 6, with $\mu = \kappa$, we obtain

$$g(\xi,\omega) = \frac{\sin\kappa\pi}{\pi}\int_0^\xi (\omega-\tau)^{\kappa-1}\,d\tau\int_0^\tau \frac{dC^\kappa(u)}{du}(\tau-u)^{-\kappa}\,du$$

$$= \int_0^\xi \frac{dC^\kappa(u)}{du}h(u)\,du \quad \dots\dots\dots\dots (1),$$

* It is also easy to prove the lemma by means of the expression of $C^\kappa(u)$ in terms of $c_1, c_2, \dots$.

† See, for example, de la Vallée-Poussin, *Cours d'analyse infinitésimale*, Vol. 2, ed. 2, Ch. 2.

‡ See the last footnote.

where
$$h(u)=\frac{\sin\kappa\pi}{\pi}\int_u^\xi(\omega-\tau)^{\kappa-1}(\tau-u)^{-\kappa}\,d\tau$$
$$=1-\frac{\sin\kappa\pi}{\pi}\int_\xi^\omega(\omega-\tau)^{\kappa-1}(\tau-u)^{-\kappa}\,d\tau.$$

Now if τ has any fixed value between ξ and ω, and $0\leqslant u\leqslant\xi$, then $(\tau-u)^{-\kappa}$ increases with u. Hence $h(u)$ is a positive decreasing function of u, always less than 1. Applying the second mean value theorem to the integral (1), we obtain

$$g(\xi,\omega)=h(0)\,C^\kappa(\eta)\qquad(0\leqq\eta\leqq\xi),$$

and this proves the result of the lemma*. In the same way we can prove

LEMMA 8. *If c_n is real, $0\leqq\xi\leqq\omega$, $\mu>0$, and $0<\kappa\leqq1$, then*
$$\frac{\Gamma(\kappa+\mu+1)}{\Gamma(\mu+1)\Gamma(\kappa)}\left|\int_0^\xi C^\mu(\tau)(\omega-\tau)^{\kappa-1}\,d\tau\right|\leqq\operatorname*{Max}_{0\leqq\tau\leqq\xi}|C^{\kappa+\mu}(\tau)|.$$

We would recommend the reader to pay close attention to Lemmas 7 and 8, and in particular the former. We shall appeal to these lemmas repeatedly in what follows, the part which they play in the theory of integrals of the 'Liouville-Riemann' type being analogous to that played by the classical mean value theorems in the theory of ordinary differential coefficients.

It should be observed that, in the proof of Lemma 7, no appeal is made to the particular structure of the function $C(\tau)$: an analogous result holds for any function which possesses an absolutely convergent integral over any finite interval. Further, the result may be extended to apply to *complex* functions; but as we shall only make use of it when $C(\tau)$ is real, we have given a proof which applies to this case only.

3. First theorem of consistency.

We can now prove

THEOREM 16. *If the series Σc_n is summable (λ,κ) to sum C, it is summable (λ,κ') to the same sum, for every κ' greater than κ.*

Writing $\kappa'=\kappa+\mu$, and applying Lemma 6, we obtain
$$C^{\kappa+\mu}(\omega)=\frac{\Gamma(\mu+\kappa+1)}{\Gamma(\kappa+1)\Gamma(\mu)}\int_0^\omega C^\kappa(u)(\omega-u)^{\mu-1}\,du.$$

But $C^\kappa(u)\sim Cu^\kappa$. Hence the theorem follows immediately from Lemma 5. In particular a convergent series is summable (λ,κ) for all positive values of κ.

* The argument is that used by Riesz, **4**. A far-reaching generalization of Lemma 7 is to be found in Riesz, **7**.

The following proposition, which we shall have occasion to use later on, is easily established by the same kind of argument: *if*

$$e^{-\rho\omega} A(\omega) = O(1),$$

then
$$e^{-(\rho+\delta)\omega} A^{\kappa'}(\omega) = o(1)$$

for any positive δ *and any* κ' *greater than* κ.

It is important to observe that *the theorem of consistency holds also for integrals* (IV, § 3). The proof is practically the same as for series.

4. Second theorem of consistency. Theorem 16 states a relation between methods of summation of the same type λ and of different orders κ, κ'. There is a much deeper theorem which concerns methods of the same order but of different types.

THEOREM 17. *If the series* Σc_n *is summable* (l, κ), *where* $l_n = e^{\lambda_n}$, *then it is summable* (λ, κ) *to the same sum.*

The proof of this theorem is somewhat intricate, and we shall confine ourselves, for the sake of simplicity, to two cases, viz. (i) that in which κ is *integral*, (ii) that in which $0 < \kappa < 1$.* These are the cases of greatest interest; and this course is one which we shall adopt in regard to a number of the theorems which follow. Further, it is easy to see that we may without loss of generality suppose C, the sum of the series, to be zero: this can always be secured by an alteration in the first term of the series.

We are given that

$$\int_1^w C_l(t)\,(w-t)^{\kappa-1}\,dt = o(w^{\kappa}) \qquad \text{.....................(1)},$$

and we have to prove that

$$\int_0^\omega C_\lambda(\tau)\,(\omega-\tau)^{\kappa-1}\,d\tau = o(\omega^{\kappa}) \qquad \text{..................(2)}.$$

If we put $\omega = \log w$, $\tau = \log t$, and observe that $C_\lambda(\log t) = C_l(t)$, we see that (2) may be written in the form

$$\int_1^w C_l(t)\,(\log w - \log t)^{\kappa-1}\frac{dt}{t} = o(\log w)^{\kappa} \qquad \text{.........(3)}.$$

* We shall indicate summarily (see § 7) the lines of the proof in the most general case.

5. (*a*) *Proof when* κ *is an integer.* In this case

$$C_l(t) = \frac{1}{\kappa!}\left(\frac{d}{dt}\right)^{\kappa} C_l^{\kappa}(t).$$

We substitute this expression for $C_l(t)$ in (3), and integrate κ times by parts. We can then show that both the terms integrated out and the integral which remains are of the form required.

In the first place, all the integrated terms vanish except one, which arises from the last integration by parts and is a constant multiple of

$$C_l^{\kappa}(w)\left[\left(\frac{d}{dt}\right)^{\kappa-1}\frac{(\log w - \log t)^{\kappa-1}}{t}\right]_{t=w}$$
$$= (-1)^{\kappa-1}(\kappa-1)!\, w^{-\kappa} C_l^{\kappa}(w) = o(1).$$

Thus we need only consider the residuary integral, which is a constant multiple of

$$\int_1^w C_l^{\kappa}(t)\left(\frac{d}{dt}\right)^{\kappa}\frac{(\log w - \log t)^{\kappa-1}}{t}\, dt \quad\text{..............(4).}$$

Now it is easily verified that

$$\left(\frac{d}{dt}\right)^{\kappa}\frac{(\log w - \log t)^{\kappa-1}}{t} = t^{-\kappa-1}\,\Sigma\, H_{r,s}(\log w)^r(\log t)^s \quad\text{...(5),}$$

where $H_{r,s}$ is a constant, and

$$r + s \leqq \kappa - 1 \quad\text{..............................(6).}$$

But, using Lemma 4 and the inequality (6), we have

$$(\log w)^r \int_1^w C_l^{\kappa}(t)(\log t)^s t^{-\kappa-1}\, dt = (\log w)^r \int_1^w (\log t)^s o\left(\frac{1}{t}\right) dt$$
$$= o(\log w)^{r+s+1} = o(\log w)^{\kappa}.$$

Hence the integral (4) is of the form required, and the theorem is proved when κ is an integer.

6. (*b*) *Proof when* $0 < \kappa < 1$. In this proof we shall suppose the *c*'s real. There is plainly no loss of generality involved in this hypothesis, as we can consider the real and imaginary parts of the series separately.

We have again to establish equation (3) of § 4. By Theorem 16, $C_l^1(t) = o(t)$. Hence we can choose ν so that

$$|C_l^1(t)| < \epsilon t \qquad (t \geqq \nu),$$

and evidently we may suppose $\nu > 1$. We then choose a value of w

greater than 3ν, and denote by M the upper limit of $C_l^1(t)$ in $(1, \nu)$; and we write the integral (3) of § 4 in the form

$$\int_1^\nu + \int_\nu^{w/3} + \int_{w/3}^w = J_1 + J_2 + J_3.$$

In the first place

$$|J_1| < M\nu\left(\log\frac{w}{\nu}\right)^{\kappa-1} = o(\log w)^\kappa \text{ ....(1).}$$

Secondly, integrating by parts, we obtain

$$J_2 = \frac{3}{w}(\log 3)^{\kappa-1} C_l^1(\tfrac{1}{3}w) - \frac{1}{\nu}\left(\log\frac{w}{\nu}\right)^{\kappa-1} C_l^1(\nu)$$

$$+ \int_\nu^{w/3} C_l^1(t)\left\{(\kappa-1)\left(\log\frac{w}{t}\right)^{\kappa-2} + \left(\log\frac{w}{t}\right)^{\kappa-1}\right\}\frac{dt}{t^2}.$$

The first two terms are in absolute value less than a constant multiple of ϵ, and the last than*

$$\epsilon\int_\nu^{w/3}\left(\log\frac{w}{t}\right)^{\kappa-1}\frac{dt}{t} < \frac{\epsilon}{\kappa}\left(\log\frac{w}{\nu}\right)^\kappa < \frac{\epsilon}{\kappa}(\log w)^\kappa.$$

Hence, for sufficiently large values of w, we have

$$|J_2| < \frac{2\epsilon}{\kappa}(\log w)^\kappa \text{ ....(2).}$$

Finally, by the second mean value theorem,

$$J_3 = \frac{3}{w}\int_{w/3}^{\xi} C_l(t)\left(\log\frac{w}{t}\right)^{\kappa-1} dt$$

$$= \frac{3}{w}\int_{w/3}^{\xi} C_l(t)(w-t)^{\kappa-1}\left(\frac{\log w - \log t}{w-t}\right)^{\kappa-1} dt,$$

where $\frac{1}{3}w \leq \xi \leq w$. Now it will easily be verified that the function

$$\left(\frac{\log w - \log t}{w-t}\right)^{\kappa-1}$$

increases steadily from $t=1$ to $t=w$, and that its limit when $t \to w$ is $w^{1-\kappa}$. Hence, using the second mean value theorem again, we obtain

$$|J_3| = \frac{3}{w}\left(\frac{\log w - \log\xi}{w-\xi}\right)^{\kappa-1}\left|\int_{\xi_1}^{\xi} C_l(t)(w-t)^{\kappa-1} dt\right|$$

$$\leq 3w^{-\kappa}\left|\int_{\xi_1}^{\xi} C_l(t)(w-t)^{\kappa-1} dt\right|,$$

* Here we use the facts that $0 < 1-\kappa < 1$ and that, as $\log 3 > 1$,

$$\left(\log\frac{w}{t}\right)^{\kappa-2} < \left(\log\frac{w}{t}\right)^{\kappa-1}$$

for $\nu < t < \frac{1}{3}w$.

where $\frac{1}{3}w \leqq \xi_1 \leqq \xi \leqq w$. But, by Lemma 7,

$$\left| \int_{\xi_1}^{\xi} C_l(t)(w-t)^{\kappa-1}\, dt \right| \leqq \frac{2}{\kappa} \operatorname*{Max}_{\frac{1}{3}w \leqq t \leqq w} |C_l^{\kappa}(t)|.$$

Hence, as $C_l^{\kappa}(t) = o(t^{\kappa})$, we have

$$J_3 = o(1) = o(\log w)^{\kappa} \quad \text{..........................(3).}$$

From (1), (2), and (3) the result of the theorem follows.

7. We have thus proved Theorem 17 when κ is an integer and when $0 < \kappa < 1$. If κ is non-integral, but greater than 1, it is necessary to combine our two methods of demonstration. We write $\mathbf{k} = [\kappa]$, and integrate the integral (3) of § 4 by parts until we have replaced $C_l(t)$ by $C_l^{\mathbf{k}}(t)$, which then plays in the proof the part played by $C_l(t)$ in § 6.

A particularly interesting special case of Theorem 17 is

THEOREM 18. *If a series is summable by arithmetic means, it is summable by logarithmic means of the same order**.

That the converse is not true is shown by the first example of IV, § 4 (6). The examples there given suggest as a general conclusion that *the efficacy of the method* (λ, κ) *increases as the rate of increase of the function* λ *decreases.* This general idea may be made more precise by the following theorem, which includes Theorem 17 as a special case, and may be established by reasoning of the same character.

THEOREM 19. *Let* μ *be any logarithmico-exponential function of* λ, *which tends to infinity with* λ, *but more slowly than* λ. *Then, if the series* Σc_n *is summable* (λ, κ), *it is summable* (μ, κ)†.

Thus, if we imagine λ as running through the functions of the logarithmico-exponential scale of infinity, such as e^n, n, $\log n$, $\log\log n$, ...‡, we obtain a sequence of systems of methods (λ, κ) of gradually increasing efficacy.

8. THEOREM 20. *If* $\lambda_1 > 0$, *and* Σc_n *is summable* (λ, κ), *then* $\Sigma c_n \lambda_n^{-\kappa}$ *is summable* (l, κ).

This theorem is interesting as a companion to Theorem 17. Its proof is very similar, though slightly more complicated. We shall suppose as before that $C=0$.

* This theorem was published without proof by Riesz, **2**.

† We can in reality say rather more, viz. that summability (λ, κ) implies summability (μ, κ) if $\mu = O(\lambda^{\Delta})$, where Δ is any constant however large. A special case of this theorem has been proved by Berwald, **1**.

‡ The result of IV, § 4, (3) shows that it is useless to consider types higher than e^n.

We are given that

$$\kappa \int_{\lambda_1}^{\omega} C_\lambda(\tau)(\omega-\tau)^{\kappa-1}\,d\tau = o(\omega^\kappa) \quad \text{.....................(1)};$$

and we have to show that

$$\kappa w^{-\kappa}\int_{l_1}^{w} D_l(t)(w-t)^{\kappa-1}\,dt = \kappa e^{-\kappa\omega}\int_{\lambda_1}^{\omega} D_\lambda(\tau)(e^\omega-e^\tau)^{\kappa-1}e^\tau\,d\tau \quad \text{...(2)}$$

tends to a limit as $\omega \to \infty$. Here $D_l(t)$, $D_\lambda(\tau)$ denote sum-functions formed from the series Σd_n, where $d_n = c_n\lambda_n^{-\kappa}$. It will easily be verified that

$$D_\lambda(\tau) = \kappa\int_{\lambda_1}^{\tau} C_\lambda(u)\frac{du}{u^{\kappa+1}} + \frac{C_\lambda(\tau)}{\tau^\kappa}.$$

We substitute this expression for $D_\lambda(\tau)$ in (2), and so obtain

$$\kappa^2 e^{-\kappa\omega}\int_{\lambda_1}^{\omega}(e^\omega-e^\tau)^{\kappa-1}e^\tau\,d\tau\int_{\lambda_1}^{\tau} C_\lambda(u)\frac{du}{u^{\kappa+1}} + \kappa e^{-\kappa\omega}\int_{\lambda_1}^{\omega}(e^\omega-e^\tau)^{\kappa-1}e^\tau C_\lambda(\tau)\frac{d\tau}{\tau^\kappa}$$

......(3).

The first term, when we invert the order of integration, and perform the integration with respect to τ, becomes

$$\kappa e^{-\kappa\omega}\int_{\lambda_1}^{\omega} C_\lambda(u)(e^\omega-e^u)^\kappa\frac{du}{u^{\kappa+1}}.$$

(*a*) We suppose first that κ is an integer. We integrate κ times by parts, as in the proof of Theorem 17. All the integrated terms vanish, so that we obtain

$$\frac{(-1)^\kappa e^{-\kappa\omega}}{(\kappa-1)!}\int_{\lambda_1}^{\omega} C_\lambda^\kappa(u)\left(\frac{d}{du}\right)^\kappa\left\{\frac{(e^\omega-e^u)^\kappa}{u^{\kappa+1}}\right\}du \quad \text{...............(4)}.$$

It will be easily verified that the differential coefficient may be expressed in the form

$$(-1)^\kappa(\kappa+1)(\kappa+2)\ldots 2\kappa\, u^{-2\kappa-1}e^{\kappa\omega} + \Sigma G_{r,s}e^{r\omega}e^{(\kappa-r)u}u^{-s} \quad \text{......(5)},$$

where $G_{r,s}$ is a constant, and

$$\kappa - r \geqq 1,\quad s-\kappa \geqq 1 \quad \text{.............................(6)}.$$

If we substitute this expression in (4), and observe that

$$u^{-s}C_\lambda^\kappa(u) = o(u^{\kappa-s}) = o(1),$$

we find that the coefficient of $G_{r,s}$ is

$$e^{-\omega(\kappa-r)}\int_{\lambda_1}^{\omega} o\{e^{(\kappa-r)u}\}\,du = o(1).$$

Hence all these terms may be neglected, and it appears that the expression (4) tends, as $\omega\to\infty$, to the limit

$$\frac{2\kappa!}{\kappa!(\kappa-1)!}\int_{\lambda}^{\infty} C_\lambda^\kappa(u)\frac{du}{u^{2\kappa+1}} \quad \text{..........................(7)}.$$

There remains the second term of (3), which may be discussed in the same manner. In this case, when we perform the integration by parts, there is one integrated term which does not vanish, as in § 5. It is however easily seen that this term has the limit zero*. The integral which remains can then be divided into a number of parts all of which can be shown to tend to zero by an argument practically identical with that employed above. Thus our final conclusion is that $\Sigma c_n \lambda_n^{-\kappa}$ is summable (l, κ), and that its sum is given by the integral (7).

We have supposed $C=0$. In order to extend our result to the general case, we have only to show that the sum of the series is given by (7) in the particular case when $c_1=C$, $c_2=c_3=\ldots=0$. This we may leave as an exercise for the reader. Finally we may observe that the theorem gives us the maximum of information possible. This may be seen by considering the case in which $\lambda_n=n$, $l_n=e^n$, $\kappa=1$. Then summability (l, κ) is equivalent to convergence, and the theorem asserts that *if* Σc_n *is summable by Cesàro's first mean, then* $\Sigma (c_n/n)$ *is convergent.* In this proposition the factor $1/n$ cannot be replaced by any factor which tends more slowly to zero.

(*b*) Suppose next that $0<\kappa<1$. As in § 6, we suppose the c's real. We have again to show that the expression (2) tends to a limit as $\omega \to \infty$. By Theorem 16, $C_\lambda^1(\tau)=o(\tau)$; and as

$$\kappa \int_{\lambda_1}^{\omega} C_\lambda(\tau) \frac{d\tau}{\tau^{\kappa+1}} = \frac{\kappa C_\lambda^1(\omega)}{\omega^{\kappa+1}} + \kappa(\kappa+1) \int_{\lambda_1}^{\omega} \frac{C_\lambda^1(\tau)}{\tau^{\kappa+2}} d\tau,$$

it follows that the integral

$$I = \kappa \int_{\lambda_1}^{\infty} C_\lambda(\tau) \frac{d\tau}{\tau^{\kappa+1}}$$

is convergent. And from this it follows, by the analogue of Theorem 16 for integrals, that

$$\kappa e^{-\kappa\omega} \int_{\lambda_1}^{\omega} C_\lambda(\tau)(e^\omega - e^\tau)^\kappa \frac{d\tau}{\tau^{\kappa+1}} \to I$$

as $\omega \to \infty$.

It remains only to show that the second term of (3) tends to zero. We separate it into two parts corresponding to the ranges of integration $(\lambda_1, 1)$, $(1, \omega)$†; and it is evident that the first part tends to zero. The second may be written in the form

$$J = \kappa e^{-\kappa\omega} \int_1^{\omega} \chi(\tau)\, C_\lambda(\tau)\, (\omega-\tau)^{\kappa-1}\, d\tau,$$

where

$$\chi(\tau) = \left(\frac{e^\omega - e^\tau}{\omega - \tau}\right)^{\kappa-1} \frac{e^\tau}{\tau^\kappa}.$$

Now it may easily be verified that, as τ increases from 1 to ω, $\chi(\tau)$ increases towards the limit $\omega^{-\kappa} e^{\kappa\omega}$. Hence, by the second mean value theorem, we obtain

$$J = \kappa\omega^{-\kappa} \int_{\xi}^{\omega} C_\lambda(\tau)(\omega-\tau)^{\kappa-1}\, d\tau, \qquad (0 \leqq \xi \leqq \omega);$$

* It is owing to the presence of this term that the series $\Sigma c_n \lambda_n^{-\kappa'}$ is not necessarily summable (l, κ) for any κ' less than κ.

† If $\lambda_1 \geqq 1$, this is unnecessary.

and it follows at once from Lemma 7 and the equation (1) that the limit of J is zero.

We leave it as an exercise for the reader to show that

$$I=\frac{\Gamma(2\kappa+1)}{\Gamma(\kappa)\,\Gamma(\kappa+1)}\int_0^\infty C_\lambda^\kappa(u)\frac{du}{u^{2\kappa+1}}.$$

9. THEOREM 21. *If Σc_n is summable (λ, κ) to sum C, then*

$$C_n - C = o\left(\frac{\lambda_{n+1}}{\lambda_{n+1}-\lambda_n}\right)^\kappa.$$

The proof of this theorem is extremely simple when κ is integral. In the equation

$$C^\kappa(\omega)=\sum_{\lambda_n\leqq\omega} c_n(\omega-\lambda_n)^\kappa = C\omega^\kappa + o(\omega^\kappa)$$

we write $\omega=\lambda_n,\ \lambda_n+h,\ \lambda_n+2h,\ \ldots,\ \lambda_n+\kappa h,$

where $\kappa h=\lambda_{n+1}-\lambda_n$, and form the κ-th difference of the $\kappa+1$ equations thus obtained. Since the κ-th difference of $(\omega-\lambda_n)^\kappa$ is a numerical multiple of h^κ, the result of the theorem follows at once.

Now let us suppose that $0<\kappa<1$; and let us assume that the c_n's are real and $C=0$, as evidently we may do without loss of generality. Then

$$\int_0^{\lambda_{n+1}}(\lambda_{n+1}-\tau)^{\kappa-1}C(\tau)\,d\tau = o(\lambda_{n+1}{}^\kappa).$$

By Lemma 7, we have also

$$\int_0^{\lambda_n}(\lambda_{n+1}-\tau)^{\kappa-1}C(\tau)\,d\tau = o(\lambda_{n+1}{}^\kappa);$$

and so the same is true of the integral taken between the limits λ_n and λ_{n+1}. But $C(\tau)=C_n$ for $\lambda_n<\tau<\lambda_{n+1}$, and so

$$(\lambda_{n+1}-\lambda_n)^\kappa C_n = o(\lambda_{n+1}{}^\kappa),$$

which proves the theorem*.

A more general theorem is

THEOREM 22. *Suppose that Σc_n is summable (λ, κ) to sum C, and that $0<\kappa'\leqq\kappa$, $\lambda_n\leqq\omega\leqq\lambda_{n+1}$. Then*

$$C^{\kappa'}(\omega)-C\omega^{\kappa'} = o\left\{\frac{\lambda_n{}^\kappa}{(\lambda_n-\lambda_{n-1})^{\kappa-\kappa'}}+\frac{\lambda_{n+1}{}^\kappa}{(\lambda_{n+1}-\lambda_n)^{\kappa-\kappa'}}\right\} \quad\ldots\ldots(1).$$

If κ' is integral, then we may write simply

$$C^{\kappa'}(\omega)-C\omega^{\kappa'} = o\left\{\frac{\lambda_{n+1}{}^\kappa}{(\lambda_{n+1}-\lambda_n)^{\kappa-\kappa'}}\right\} \quad\ldots\ldots\ldots\ldots\ldots(2);$$

and this result holds for $\kappa'=0$, provided $\lambda_n<\omega$.†

* The proof for the case in which κ is non-integral and greater than 1 is contained in that of Theorem 22.

† This distinction arises from the fact that $C(\omega)$ is discontinuous for $\omega=\lambda_n$.

We shall as usual take $C=0$, and assume that c_n is real. First suppose that κ' is an integer, and let us write $\mathbf{k}=[\kappa]$.* Let us also write

$$\Omega_1=\lambda_n+h,\ \ldots,\ \Omega_{\mathbf{k}}=\lambda_n+\mathbf{k}h,\ \Omega_{\mathbf{k}+1}=\lambda_{n+1},$$

where

$$h=(\lambda_{n+1}-\lambda_n)/(\mathbf{k}+1).$$

If Ω denotes any one of these numbers, we have, by Lemma 6,

$$\frac{\Gamma(\kappa+1)}{\Gamma(\mathbf{k}+1)\Gamma(\kappa-\mathbf{k})}\int_0^{\Omega} C^{\mathbf{k}}(\tau)(\Omega-\tau)^{\kappa-\mathbf{k}-1}\,d\tau=C^{\kappa}(\Omega).$$

Using Lemma 8 and the equations $C^{\kappa}(\Omega)=o(\Omega^{\kappa})=o(\lambda_{n+1}{}^{\kappa})$, we obtain

$$\int_{\lambda_n}^{\Omega} C^{\mathbf{k}}(\tau)(\Omega-\tau)^{\kappa-\mathbf{k}-1}\,d\tau=o(\lambda_{n+1}{}^{\kappa}) \quad \text{......................(3)}.$$

Integrating $\mathbf{k}$ times by parts, and observing that $C(\tau)$ is constant in the range of integration, we can express the integral in (3) as the sum of constant multiples of the $\mathbf{k}+1$ functions

$$(\lambda_{n+1}-\lambda_n)^{\kappa-\mathbf{k}}C^{\mathbf{k}}(\lambda_n),\ (\lambda_{n+1}-\lambda_n)^{\kappa-\mathbf{k}+1}C^{\mathbf{k}-1}(\lambda_n),$$
$$\ldots,\ (\lambda_{n+1}-\lambda_n)^{\kappa}C(\lambda_n+0) \quad \text{.........(4)†}.$$

This process leads, for the $\mathbf{k}+1$ different values of Ω, to $\mathbf{k}+1$ different linear combinations of the functions (4), each of which is of the form $o(\lambda_{n+1}{}^{\kappa})$. But it is easy to verify that these $\mathbf{k}+1$ linear combinations are linearly independent, the determinant of the system being a 'Vandermonde-Cauchy'‡ determinant, different from zero; and so the functions themselves are of this form.

The last function is $(\lambda_{n+1}-\lambda_n)^{\kappa}C(\omega)$, and so the theorem is proved for $\kappa'=0$. The last function but one is $(\lambda_{n+1}-\lambda_n)^{\kappa-1}C^1(\lambda_n)$; and so $C^1(\lambda_n)$ is of the form prescribed by the theorem. Hence

$$C^1(\omega)=C^1(\lambda_n)+\int_{\lambda_n}^{\omega}C(\tau)\,d\tau=C^1(\lambda_n)+\int_{\lambda_n}^{\omega}o\left\{\frac{\lambda_{n+1}{}^{\kappa}}{(\lambda_{n+1}-\lambda_n)^{\kappa}}\right\}d\tau$$
$$=o\left\{\frac{\lambda_{n+1}{}^{\kappa}}{(\lambda_{n+1}-\lambda_n)^{\kappa-1}}\right\}.$$

Hence the theorem is proved for $\kappa'=1$. Repeating the argument we establish it for $\kappa'=0, 1, 2, \ldots, \mathbf{k}$.

We pass now to the case in which κ' is not integral, and we write $\mathbf{k}'=[\kappa']$.§

* We shall in this case give a complete proof for all values of κ, integral or not. The method of proof is that used by Riesz, **4**, in proving the equivalence of the 'arithmetic' means with Cesàro's.

† In the last function we must write λ_n+0 and not λ_n, on account of the discontinuity of $C(\tau)$.

‡ See Pascal, *I Determinanti* (Manuali Hoepli), pp. 166 *et seq.*

§ It is very curious that the simpler result which holds when κ' is integral should not hold always; but it is possible to show by examples that this is so.

By Lemma 6, we have

$$C^{\kappa'}(\omega)=\frac{\Gamma(\kappa'+1)}{\Gamma(\mathbf{k}'+1)\,\Gamma(\kappa'-\mathbf{k}')}\int_0^{\omega} C^{\mathbf{k}'}(\tau)\,(\omega-\tau)^{\kappa'-\mathbf{k}'-1}\,d\tau$$

$$=\frac{\Gamma(\kappa'+1)}{\Gamma(\mathbf{k}'+1)\,\Gamma(\kappa'-\mathbf{k}')}\left(\int_0^{\lambda_{n-1}}+\int_{\lambda_{n-1}}^{\omega}\right)=J_1+J_2,$$

say. We begin by considering J_2. Dividing the interval (λ_{n-1}, ω) into the two parts $(\lambda_{n-1}, \lambda_n)$, (λ_n, ω), and using the result (2) for $C^{\mathbf{k}'}(\tau)$, we find

$$J_2=o\left\{\frac{\lambda_n^{\kappa}}{(\lambda_n-\lambda_{n-1})^{\kappa-\mathbf{k}'}}\right\}\int_{\lambda_{n-1}}^{\lambda_n}(\omega-\tau)^{\kappa'-\mathbf{k}'-1}\,d\tau$$

$$+\,o\left\{\frac{\lambda_{n+1}^{\kappa}}{(\lambda_{n+1}-\lambda_n)^{\kappa-\mathbf{k}'}}\right\}\int_{\lambda_n}^{\omega}(\omega-\tau)^{\kappa'-\mathbf{k}'-1}\,d\tau.$$

Now $\kappa'-\mathbf{k}'-1$ is negative and $\kappa'-\mathbf{k}'$ positive. Hence

$$\int_{\lambda_{n-1}}^{\lambda_n}(\omega-\tau)^{\kappa'-\mathbf{k}'-1}\,d\tau\leqq\int_{\lambda_{n-1}}^{\lambda_n}(\lambda_n-\tau)^{\kappa'-\mathbf{k}'-1}\,d\tau=\frac{(\lambda_n-\lambda_{n-1})^{\kappa'-\mathbf{k}'}}{\kappa'-\mathbf{k}'},$$

$$\int_{\lambda_n}^{\omega}(\omega-\tau)^{\kappa'-\mathbf{k}'-1}\,d\tau=\frac{(\omega-\lambda_n)^{\kappa'-\mathbf{k}'}}{\kappa'-\mathbf{k}'}\leqq\frac{(\lambda_{n+1}-\lambda_n)^{\kappa'-\mathbf{k}'}}{\kappa'-\mathbf{k}'}.$$

Thus

$$J_2=o\left\{\frac{\lambda_n^{\kappa}}{(\lambda_n-\lambda_{n-1})^{\kappa-\kappa'}}\right\}+o\left\{\frac{\lambda_{n+1}^{\kappa}}{(\lambda_{n+1}-\lambda_n)^{\kappa-\kappa'}}\right\}\quad\text{..............(5).}$$

In order to obtain an upper limit for J_1, we integrate $\mathbf{k}-\mathbf{k}'$ times by parts. We find that

$$J_1=\sum_{\mu=1}^{\mathbf{k}-\mathbf{k}'}\frac{\Gamma(\kappa'+1)}{\Gamma(\mathbf{k}'+\mu+1)\,\Gamma(\kappa'-\mathbf{k}'-\mu+1)}(\omega-\lambda_{n-1})^{\kappa'-\mathbf{k}'-\mu}\,C^{\mathbf{k}'+\mu}(\lambda_{n-1})$$

$$+\frac{\Gamma(\kappa'+1)}{\Gamma(\mathbf{k}+1)\,\Gamma(\kappa'-\mathbf{k})}\int_0^{\lambda_{n-1}}C^{\mathbf{k}}(\tau)\,(\omega-\tau)^{\kappa'-\mathbf{k}-1}\,d\tau.$$

Since $\kappa'-\mathbf{k}'-\mu<0$, we may replace the powers of $\omega-\lambda_{n-1}$ in the first line by the corresponding powers of $\lambda_n-\lambda_{n-1}$. If we do this, and at the same time apply the result (2) to the factors $C^{\mathbf{k}'+\mu}(\lambda_{n-1})$, we find at once that every term in the first line is of the form

$$o\left\{\frac{\lambda_n^{\kappa}}{(\lambda_n-\lambda_{n-1})^{\kappa-\kappa'}}\right\}\quad\text{..............................(6).}$$

On the other hand, the integral which occurs in the second line may, by the second mean value theorem, be expressed in the form

$$(\omega-\lambda_{n-1})^{\kappa'-\kappa}\int_{\xi}^{\lambda_{n-1}}C^{\mathbf{k}}(\tau)\,(\omega-\tau)^{\kappa-\mathbf{k}-1}\,d\tau,$$

where $0\leqq\xi\leqq\lambda_{n-1}$. Replacing ω by λ_n in the external factor, and applying Lemma 8 to the integral, in the same way in which we used Lemma 7 in the proof of Theorem 21, we see that this part of J_1 is also of the form (6). The proof of Theorem 22 is thus completed.

In the particularly important case in which $\lambda_n=n$, Theorem 22 shows that *if the series* Σc_n *is summable* (n, κ), *and* $0\leqq\kappa'<\kappa$, *then*

$$C^{\kappa'}(\omega)=o(\omega^{\kappa}).$$

VI

ABELIAN AND TAUBERIAN THEOREMS

1. Generalisations of Theorem 2 and its corollaries. We pass now to an important theorem which occupies the same place in the theory of the summability of Dirichlet's series as does Theorem 2 in the elementary theory of their convergence. But first we shall prove a subsidiary proposition which will be useful to us.

LEMMA 9. *If $f(u)$ is integrable, and $f(u)=o(u^\kappa)$, then*

$$\int_0^\omega s^{\kappa+1}e^{-su}f(u)\,du$$

tends to a limit as $\omega\to\infty$, uniformly for all values of s in the angle **a** *defined by $|\operatorname{am} s|\leqq\alpha<\frac{1}{2}\pi$.*

Choose ξ so that

$$|f(u)|<\epsilon u^\kappa \qquad (u\geqq\xi).$$

Then

$$\left|\int_{\omega_1}^{\omega_2} s^{\kappa+1}e^{-su}f(u)\,du\right|<\epsilon\int_0^\infty |s|^{\kappa+1}e^{-\sigma u}u^\kappa\,du$$

$$=\epsilon\Gamma(\kappa+1)\left(\frac{|s|}{\sigma}\right)^{\kappa+1}\leqq\epsilon\Gamma(\kappa+1)(\sec\alpha)^{\kappa+1},$$

if only $\omega_2>\omega_1\geqq\xi$. As ξ is independent of the position of s in the angle, the lemma is proved*.

2. THEOREM 23. *If Σa_n is summable (λ,κ), then $\Sigma a_n e^{-\lambda_n s}$ is uniformly summable throughout the angle* **a**.

THEOREM 24. *The sum of the series $\Sigma a_n e^{-\lambda_n s}$ is equal to*

$$\frac{1}{\Gamma(\kappa+1)}\int_0^\infty s^{\kappa+1}e^{-s\tau}A^\kappa(\tau)\,d\tau$$

at all points of **a** *other than the origin; and to*

$$A+\frac{1}{\Gamma(\kappa+1)}\int_0^\infty s^{\kappa+1}e^{-s\tau}\{A^\kappa(\tau)-A\tau^\kappa\}\,d\tau,$$

where A is the sum of Σa_n, at all points of **a**.†

* The presence of the factor $s^{\kappa+1}$ is of course essential for the truth of the result.

† Compare IV, § 4, (1) for the simplest case of such a representation of a Dirichlet's series by an integral.

We shall, as usual, prove this in the cases in which (i) κ is integral and (ii) $0<\kappa<1$. We observe first that, if $c_n=a_n e^{-\lambda_n s}$, and λ_p denotes the last λ less than ω, then

$$C^\kappa(\omega)=\sum_{\lambda_n<\omega} c_n(\omega-\lambda_n)^\kappa=\sum_1^{p-1} A_n \Delta\{e^{-\lambda_n s}(\omega-\lambda_n)^\kappa\}+A_p e^{-\lambda_p s}(\omega-\lambda_p)^\kappa$$

$$=-\int_0^\omega A(\tau)\frac{d}{d\tau}\{e^{-s\tau}(\omega-\tau)^\kappa\}\,d\tau \quad \text{.... ......(1).}$$

(*a*) *Proof when κ is integral.* We suppose first that

$$A=0,\ A^\kappa(\omega)=o(\omega^\kappa).$$

Integrate (1) κ times by parts. All the integrated terms vanish save one, which is

$$e^{-s\omega}A^\kappa(\omega)=o(\omega^\kappa) \quad \text{......................(2),}$$

uniformly throughout $\mathbf{a}$. Thus we need only consider the integral remaining over, which, when divided by ω^κ, is

$$\frac{(-1)^{\kappa+1}\omega^{-\kappa}}{\kappa!}\int_0^\omega A^\kappa(\tau)\left(\frac{d}{d\tau}\right)^{\kappa+1}\{e^{-s\tau}(\omega-\tau)^\kappa\}\,d\tau \quad \text{.........(3).}$$

Now it will easily be verified that

$$\left(\frac{d}{d\tau}\right)^{\kappa+1}\{e^{-s\tau}(\omega-\tau)^\kappa\}=(-1)^{\kappa+1}e^{-s\tau}s^{\kappa+1}\omega^\kappa+e^{-s\tau}\Sigma H_{i,j,k}s^i\omega^j\tau^k,$$

where $H_{i,j,k}$ is a constant, and

$$i=j+k+1,\ j\leqq\kappa-1 \quad \text{......................(4).}$$

When we substitute this expression in (3), the first term, which we may call the *principal* term, gives rise to the integral

$$\frac{1}{\kappa!}\int_0^\omega s^{\kappa+1}e^{-s\tau}A^\kappa(\tau)\,d\tau \quad \text{......................(5).}$$

As $\omega\to\infty$, this integral, by Lemma 9, tends uniformly to a limit, viz. the first integral of Theorem 24. Thus, when $A=0$, all that is necessary to complete the proof is to show that

$$\omega^{-\kappa+j}\int_0^\omega s^i\tau^k e^{-s\tau}A^\kappa(\tau)\,d\tau=o(1),$$

uniformly in $\mathbf{a}$; i, j, k being subject to the inequalities (4). We divide this integral into the two parts

$$\omega^{-\kappa+j}\int_0^\nu+\omega^{-\kappa+j}\int_\nu^\omega=J_1+J_2,$$

choosing ν so that $|A^\kappa(\tau)|<\epsilon\tau^\kappa$ throughout the range of integration in J_2. Further we observe that there is a constant M such that $|A^\kappa(\tau)|<M\tau^\kappa$ for all values of τ.

The function

$$F(x) = x^{-\kappa+j} \int_0^x e^{-\tau} \tau^{k+\kappa} \, d\tau$$

has (for positive values of x) a maximum μ.* And

$$|J_2| < \epsilon\omega^{-\kappa+j} |s|^i \int_0^\omega e^{-\sigma\tau} \tau^{k+\kappa} \, d\tau$$

$$= \epsilon\omega^{-\kappa+j} \left(\frac{|s|}{\sigma}\right)^i \sigma^{i-k-\kappa-1} \int_0^{\sigma\omega} e^{-y} y^{k+\kappa} \, dy$$

$$= \epsilon (|s|/\sigma)^i F(\sigma\omega) \dagger \leqq \epsilon\mu (\sec a)^i \quad \text{.........(6).}$$

Also

$$|J_1| < M\omega^{-\kappa+j} |s|^i \int_0^\nu e^{-\sigma\tau} \tau^{k+\kappa} \, d\tau$$

$$= M\omega^{-\kappa+j} (|s|/\sigma)^i \nu^{\kappa-j} F(\nu\sigma)$$

$$\leqq M\mu (\sec a)^i (\nu/\omega)^{\kappa-j} \quad \text{.........................(7).}$$

From (6) and (7) it follows that, by taking first ν and then ω sufficiently large, we can make $J_1 + J_2$ as small as we please. This completes the proof of Theorems 23 and 24, when κ is an integer and $A = 0$.

Now suppose $A \neq 0$. Then the series

$$(-A + a_1 e^{-\lambda_1 s}) + a_2 e^{-\lambda_2 s} + a_3 e^{-\lambda_3 s} + \dots$$

is, in virtue of what has just been proved, uniformly summable in **a**, with sum

$$\frac{1}{\kappa!} \int_0^\infty s^{\kappa+1} e^{-s\tau} \{A^\kappa(\tau) - A\tau^\kappa\} \, d\tau. \ddagger$$

And the series

$$A + 0 + 0 + \dots$$

is, as may be seen at once by actual formation of the typical means, or inferred from the theorem of consistency, also uniformly summable in **a**. Moreover its sum is A, which, *except when* $s = 0$, is equal to

$$\frac{A}{\kappa!} \int_0^\infty s^{\kappa+1} e^{-s\tau} \tau^\kappa \, d\tau.$$

Combining these results we obtain Theorems 23 and 24. It should be observed that, when $A \neq 0$, the first integral in Theorem 24 is *not* uniformly convergent, or even continuous for $s = 0$.

* Since $$-\kappa + j + k + \kappa + 1 = j + k + 1 = i > 0,$$
the function has the limit 0 when $x \to 0$ as well as when $x \to \infty$.

† Since $$i - k - \kappa - 1 = -\kappa + j.$$

‡ The sum function of the modified series is equal to $A(\tau) - A$.

3. (*b*) *Proof when* $0 < \kappa < 1$. As before, we begin by supposing $A = 0$. We have, by (1) of § 2, to discuss the limit of

$$-\omega^{-\kappa} \int_0^\omega A(\tau) \frac{d}{d\tau} \{e^{-s\tau} (\omega - \tau)^\kappa\} d\tau = \kappa\omega^{-\kappa} e^{-s\omega} \int_0^\omega A(\tau)(\omega - \tau)^{\kappa-1} d\tau$$
$$- \omega^{-\kappa} \int_0^\omega A(\tau) \frac{d}{d\tau} \{(e^{-s\tau} - e^{-s\omega})(\omega - \tau)^\kappa\} d\tau.$$

The first term on the right-hand side is

$$e^{-s\omega} \omega^{-\kappa} A^\kappa(\omega) = o(1)$$

uniformly in **a**. Thus we need only consider the second term, which, when we integrate by parts, takes the form

$$\omega^{-\kappa} \int_0^\omega A^1(\tau) \frac{d^2}{d\tau^2} \{(e^{-s\tau} - e^{-s\omega})(\omega - \tau)^\kappa\} d\tau = J_1 + J_2 + J_3 \quad \ldots(1),$$

where J_1, J_2, J_3 are three integrals containing, under the sign of integration, the function $A^1(\tau)$ multiplied respectively by

$$\left.\begin{array}{ll} \text{(i)} & s^2 e^{-s\tau} (\omega - \tau)^\kappa, \\ \text{(ii)} & 2\kappa s e^{-s\tau} (\omega - \tau)^{\kappa-1}, \\ \text{(iii)} & \kappa(\kappa - 1)(e^{-s\tau} - e^{-s\omega})(\omega - \tau)^{\kappa-2} \end{array}\right\} \quad \ldots\ldots\ldots\ldots(2).$$

In the first place

$$J_1 = \omega^{-\kappa} \int_0^\omega s^2 e^{-s\tau} A^1(\tau)(\omega - \tau)^\kappa d\tau \quad \ldots\ldots\ldots\ldots\ldots(3).$$

Now Σa_n is summable $(\lambda, 1)$, and so $A^1(\tau) = o(\tau)$. Hence, by Lemma 9, the integral

$$\int_0^\omega s^2 e^{-s\tau} A^1(\tau) d\tau$$

tends to a limit, as $\omega \to \infty$, uniformly in **a**. And hence, by the analogue for integrals* of Theorem 16, the integral (3) does the same. Further, the value of the limit is

$$\int_0^\infty s^2 e^{-s\tau} A^1(\tau) d\tau = \frac{s^2}{\Gamma(1+\kappa)\Gamma(1-\kappa)} \int_0^\infty e^{-s\tau} d\tau \int_0^\tau A^\kappa(u) \frac{du}{(\tau - u)^\kappa}$$
$$= \frac{s^2}{\Gamma(1+\kappa)\Gamma(1-\kappa)} \int_0^\infty A^\kappa(u) du \int_u^\infty e^{-s\tau} \frac{d\tau}{(\tau - u)^\kappa}$$
$$= \frac{1}{\Gamma(1+\kappa)} \int_0^\infty s^{\kappa+1} e^{-su} A^\kappa(u) du. \dagger$$

* See the end of V, § 3.

† In the first line we use Lemma 6. The inversion of the order of integration presents no difficulty, all the integrals concerned being absolutely convergent.

It remains to prove that J_2 and J_3 tend uniformly to zero. We write

$$J_2 = 2\kappa\omega^{-\kappa}\int_0^\omega se^{-s\tau}(\omega-\tau)^{\kappa-1}A^1(\tau)\,d\tau = 2\kappa\omega^{-\kappa}\left(\int_0^\nu + \int_\nu^\omega\right) = J_2' + J_2'',$$

say. We can choose ν so that $|A^1(\tau)| < \epsilon\tau$ for $\tau > \nu$, and we can suppose $\omega - \nu > 1$. Further, there is a constant M such that $|A^1(\tau)| < M\tau$ for all positive values of τ. Also $\tau e^{-\tau} < 1$ for all values of τ, and $|s|/\sigma \leqq \sec\alpha$ throughout the angle **a**. Hence, denoting by K the constant $2\kappa\sec\alpha$, we have

$$|J_2'| < KM\omega^{-\kappa}\int_0^\nu \sigma\tau e^{-\sigma\tau}(\omega-\tau)^{\kappa-1}\,d\tau < KM\nu\omega^{-\kappa} \quad \ldots\ldots(4).$$

Similarly we have

$$|J_2''| < K\epsilon\omega^{-\kappa}\int_\nu^\omega \sigma\tau e^{-\sigma\tau}(\omega-\tau)^{\kappa-1}\,d\tau < \frac{K\epsilon}{\kappa} \quad \ldots\ldots\ldots(5).$$

From (4) and (5) it follows that we can make J_2 as small as we please, uniformly throughout **a**, by making first ν and then ω sufficiently large.

In order to discuss J_3 we observe that, by Lemma 2,

$$|e^{-s\tau} - e^{-s\omega}| \leqq (e^{-\sigma\tau} - e^{-\sigma\omega})\sec\alpha < (\omega-\tau)\,\sigma e^{-\sigma\tau}\sec\alpha.$$

The discussion is then almost exactly the same as in the case of J_2. The proof of the theorems is thus completed.

4. Lines of summability. Analytic character of the sum.

From Theorem 23 we can at once deduce a series of important corollaries, analogous to those deduced from Theorem 2 in II, §§ 2 *et seq.*

THEOREM 25. *If the series is summable* (λ, κ)* *for a value of* s *whose real part is* σ, *then it is summable* (λ, κ) *for all values of* s *whose real part is greater than* σ.

THEOREM 26. *There is a number* σ_κ *such that the series is summable when* $\sigma > \sigma_\kappa$ *and not summable when* $\sigma < \sigma_\kappa$. *We may have* $\sigma_\kappa = -\infty$ *or* $\sigma_\kappa = \infty$.†

We now define the abscissa σ_κ, the line $\sigma = \sigma_\kappa$, and the half-plane $\sigma > \sigma_\kappa$ of summability (λ, κ), just as we did in II, § 2 when $\kappa = 0$. It is evident (from the first theorem of consistency) that

$$\bar{\sigma} \geqq \sigma_0 \geqq \sigma_1 \geqq \sigma_2 \geqq \ldots.$$

* It should be added that the result of Theorem 25 remains true if we assume only that Σa_n is *finite* (λ, κ): cf. the first footnote to p. 4. The first representation of the sum as an integral is also valid in this case, as may easily be shown by a trifling modification of the proof of Theorem 24.

† See p. 4 for an explanation of the meaning to be attached to this phrase.

THEOREM 27. *If D is any finite region for all points of which $\sigma \geqq \sigma_\kappa + \delta > \sigma_\kappa$, then the series is uniformly summable (λ, κ) throughout D, and its sum represents a branch $f(s)$ of an analytic function regular throughout D. Further, the series*

$$\Sigma\, a_n \lambda_n^{\rho}\, e^{-\lambda_n s},$$

where ρ is any number real or complex, and λ_n^{ρ} has its principal value, is also uniformly summable (λ, κ) throughout D, and, when ρ is a positive integer, represents the function

$$(-1)^{\rho} f^{(\rho)}(s).$$

The proof of this theorem is similar to that of Theorem 4. One additional remark is however necessary. When we prove that the summability of Σa_n involves that of

$$\Sigma a_n \lambda_n^{\rho} = \Sigma a_n e^{\rho \log \lambda_n},$$

whenever the real part of ρ is negative, we must appeal, not to Theorem 23, but to its analogue Theorem 29 below; for the means (λ, κ) are the means of the second kind for the series $\Sigma a_n \lambda_n^{-s}$.

THEOREM 28. *If the series is summable (λ, κ) for $s = s_0$, and has the sum $f(s_0)$, then $f(s) \to f(s_0)$ when $s \to s_0$ along any path lying entirely inside the angle whose vertex is at s_0 and which is similar and similarly situated to the angle* **a**.

There is also an obvious generalisation of Theorem 6 which we shall not state at length.

5. Summability by typical means of the second kind. In §§ 1—4 we have considered exclusively typical means of the first kind. All the results of these sections, however, remain true when we work with means of the second kind, except that Theorems 23 and 24 must be replaced by

THEOREM 29. *If the series Σa_n is summable (l, κ), then the series $\Sigma a_n l_n^{-s}$ is uniformly summable (l, κ) in the angle* **a**. *Its sum is (except for $s = 0$) equal to the integral*

$$\frac{\Gamma(s+\kappa+1)}{\Gamma(\kappa+1)\Gamma(s)} \int_1^\infty A_l^\kappa(u)\, u^{-s-\kappa-1}\, du.*$$

It is not necessary that we should do more than indicate the lines of the proof when $0 < \kappa < 1$. The κ-th mean formed from $\Sigma a_n l_n^{-s}$ may be expressed, by the same transformation as was used at the beginning of § 2, in the form

$$-w^{-\kappa} \int_1^w A_l(t) \frac{d}{dt} \{t^{-s} (w-t)^\kappa\}\, dt.$$

Arguing as at the beginning of § 3, we replace this expression by

$$w^{-\kappa} \int_1^w A_l^1(t) \frac{d^2}{dt^2} \{(t^{-s} - w^{-s})(w-t)^\kappa\}\, dt.$$

* As $l_1 = e^{\lambda_1} \geqq 1$, and $A_l^\kappa(u) = 0$ for $u \leqq l_1$, the lower limit may be 0, 1, or l_1 indifferently.

Finally, as $w \to \infty$, this tends uniformly to the limit

$$s(s+1)\int_1^\infty A_l^1(t)\,t^{-s-2}\,dt = \frac{s(s+1)}{\Gamma(1+\kappa)\,\Gamma(1-\kappa)}\int_1^\infty t^{-s-2}\,dt\int_1^t A_l^\kappa(u)\frac{du}{(t-u)^\kappa}$$

$$= \frac{s(s+1)}{\Gamma(1+\kappa)\,\Gamma(1-\kappa)}\int_1^\infty A_l^\kappa(u)\,du\int_u^\infty \frac{t^{-s-2}}{(t-u)^\kappa}\,dt$$

$$= \frac{\Gamma(s+\kappa+1)}{\Gamma(\kappa+1)\,\Gamma(s)}\int_1^\infty A_l^\kappa(u)\,u^{-s-\kappa-1}\,du.$$

We add one more theorem.

THEOREM 30. *The lines of summability are the same for the means of the first and the second kind.*

We shall content ourselves with sketching the proof of this theorem. In the first place, if the series is summable (l, κ) for $s = s_0$, it is, by Theorem 17, summable (λ, κ) for $s = s_0$, and *a fortiori* for $\sigma > \sigma_0$. On the other hand, if it is summable (λ, κ) for $s = s_0$, the series

$$\Sigma\, a_n \lambda_n^{-p} e^{-\lambda_n s},$$

where p is any integer greater than κ, is, by Theorem 20, summable (l, κ) for $s = s_0$, and *a fortiori* for $\sigma > \sigma_0$. Hence, by the analogue of Theorem 27 for means of the second kind, the original series is summable (l, κ) for $\sigma > \sigma_0$.*

6. **Explicit formulae for σ_κ.** The actual values of the abscissae of summability are given by the following generalisation of Theorem 8.

THEOREM 31. *The abscissa of summability σ_κ, if positive, is given by*

$$\sigma_\kappa = \overline{\lim}\,\frac{\log|A_\lambda^\kappa(\omega)|}{\omega} = \overline{\lim}\,\frac{\log|A_l^\kappa(w)|}{\log w} - \kappa.$$

The proof of these results follows the general lines of that of Theorems 23 and 24, but is easier, as no question of uniformity is involved. As the proof is not very interesting in itself, we shall confine ourselves to indicating the general line of the argument for means of the first kind.

We assume first that

$$A^\kappa(\tau) = o\{e^{\tau(\eta+\delta)}\} \quad \text{..............................(1)}$$

for a definite positive η and every positive δ. If now we follow the argument of §§ 2, 3, we can show without difficulty that the series is summable (λ, κ) if $\sigma > \eta$, and that its sum is the first integral of Theorem 24. If on the other hand the series is summable (λ, κ) when $s = \eta + \delta$, and

$$c_n = a_n e^{-\lambda_n s},\quad a_n = c_n e^{\lambda_n s},$$

we have obviously

$$C^\kappa(\tau) = o(e^{\epsilon\tau}),$$

* It is also possible to give a direct proof of this theorem similar to, but rather easier than, that of Theorem 20. We have to prove that the summability (λ, κ) of $\Sigma\, a_n e^{-\lambda_n s}$ involves the summability (l, κ) of $\Sigma\, a_n e^{-\lambda_n (s+\delta)}$ for any positive δ.

for every positive ϵ. Hence, performing the same arguments with $-s$ in the place of s, we deduce (1) with $\delta+\epsilon$ in the place of δ. It follows that if (1) holds for η, but for no smaller number than η, then the series is summable when $\sigma>\eta$ but not when $\sigma<\eta$. This proves the first equation in Theorem 31.

7. Tauberian Theorems. In this section we shall state a number of theorems whose general character is 'Tauberian'; that is to say, which are developments of an idea which appeared first in Tauber's well-known 'converse of Abel's Theorem'*. In spite of the great intrinsic interest of these theorems we omit the proofs, as we shall not have occasion to make any applications of the results.

THEOREM 32†. *If*

$$\text{(i)}\quad a_n = o\{(\lambda_n - \lambda_{n-1})/\lambda_n\}$$

and (ii) *the series* $\sum a_n e^{-\lambda_n s}$, *then certainly convergent for* $\sigma > 0$, *tends to a limit* A *as* $s \to 0$ *through positive values, then the series* $\sum a_n$ *is convergent and has the sum* A.

THEOREM 33‡. *The conclusion of Theorem* 32 *still holds if the condition* (i) *is replaced by the more general condition*

$$\text{(i')}\quad \lambda_1 a_1 + \lambda_2 a_2 + \dots + \lambda_n a_n = o(\lambda_n).$$

Moreover the conditions (i') *and* (ii) *are necessary and sufficient for the convergence of the series* $\sum a_n$.

THEOREM 34§. *If* $\lambda_n - \lambda_{n-1} = o(\lambda_n)$, *then the condition* (i) *of Theorem* 32 *may be replaced by the more general condition*

$$a_n = O\{(\lambda_n - \lambda_{n-1})/\lambda_n\}.$$

THEOREM 35‖. *If*

$$\text{(i)}\quad a_n = O\{(\lambda_n - \lambda_{n-1})/\lambda_n\}$$

and (ii) $\sum a_n$ *is summable* (λ, κ) *to sum* A, *then* $\sum a_n$ *is convergent to sum* A.

THEOREM 36. *If*

$$\lambda_n = O(\lambda_n - \lambda_{n-1}),$$

then no series can be summable (λ, κ) *unless it is convergent.*

The last theorem is an immediate consequence of Theorem 21. It contains as a particular case the result of IV, § 4, (3); viz. that the

* Tauber, **1**. For a general explanation of the character of a 'Tauberian' theorem see Hardy and Littlewood, **1**.

† Landau, **3**. ‡ Schnee, **3**. § Littlewood, **1**. See also Landau, **11**, **12**.

‖ Hardy, **8**. If $\lambda_n - \lambda_{n-1} = o(\lambda_n)$, this may be deduced as a corollary from Theorems 28 and 34. See also Hardy, **4**.

means (e^n, κ) are 'trivial' in the sense that no non-convergent series is summable by means of them.

The theorems of this section are capable of many interesting generalisations for which we must refer elsewhere*. We add however one important theorem which resembles Theorems 32—36 in that its conditions include a condition as to the order or average order of the coefficient a_n, but differs from them fundamentally in that it depends on the theory of functions of a complex variable.

THEOREM 37†. *If*

$$\text{(i)} \quad A_n = a_1 + a_2 + \dots + a_n = o(e^{\lambda_n c}) \qquad (c \geqq 0)$$

and (ii) *the series* $\Sigma a_n e^{-\lambda_n s}$, *then certainly convergent for* $\sigma > c$, *represents a function* $f(s)$ *regular for* $s = s_0 = c + t_0 i$, *then the series is convergent for* $s = s_0$, *and its sum is* $f(s_0)$.

It should be observed that (i) is certainly satisfied if $c > 0$ and

$$\text{(i')} \quad a_n = o\{(\lambda_n - \lambda_{n-1})\, e^{\lambda_{n-1} c}\}.$$

This is no longer true if $c = 0$. But it is easy to see, by applying a linear transformation to the variable s, that the theorem obtained by putting $c = 0$ in (i'), viz. '*if*

$$a_n = o(\lambda_n - \lambda_{n-1})$$

then the series $\Sigma a_n e^{-\lambda_n s}$ *is convergent at every regular point of the line* $\sigma = 0$' is certainly true in all cases in which $\lambda_n - \lambda_{n-1} = O(1)$. This theorem is the direct generalisation of a well-known theorem of Fatou ‡, to which it reduces when $\lambda_n = n$. It should also be observed that Fatou's theorem and its extension become false when $O(\lambda_n - \lambda_{n-1})$ is substituted for $o(\lambda_n - \lambda_{n-1})$.

8. **Examples to illustrate §§ 4—7.** (1) For the series Σn^{-s}, we have

$$\bar{\sigma} = \sigma_0 = \sigma_1 = \sigma_2 = \dots = 1.$$

See IV, § 4, (6).

(2) For the series $\Sigma(-1)^n n^{-s}$, we have $\sigma_\kappa = -\kappa$. See IV, § 4, (4).

(3) For the series $\Sigma e^{Ain^a} n^{-s}$, where $0 < a < 1$ and $A \neq 0$, we have $\sigma_\kappa = 1 - (\kappa + 1)\, a$. See IV, § 4, (5).

* See in particular Landau, **8**; Hardy and Littlewood, **2**, **4**, **5**. In reference to the original theorem of Tauber see Pringsheim, **2**, **3**; Bromwich, *Infinite series*, p. 251.

† Riesz, **4**. The proof of the general theorem is still unpublished. For the case $\lambda_n = n$ see Riesz, **5**; for the case $\lambda_n = \log n$ see Landau, **8**. The condition (i) is a necessary condition for the existence of any points of convergence on the line $\sigma = c$ (Jensen, **2**).

‡ Fatou, **1**; Riesz, **5**. The latter paper contains a number of further theorems of a similar character.

(4) Each of the series (2) and (3) is summable, by typical means of *some* order, all over the plane, and consequently represents an integral function of s. It is of some interest to obtain an example of a series which represents an integral function, but cannot, for some values of s, be summed by any typical mean. Such an example is afforded by the series

$$\Sigma e^{i(\log n)^2} n^{-s}.$$

Here all the lines of summability coincide in the line $\sigma = 1$. None the less the series represents an integral function. So does the series

$$\Sigma (-1)^n e^{-sn^a} \qquad (0 < a < 1),$$

all of whose lines of summability coincide in the line $\sigma = 0$.

(5) For the series

$$1^{-s} - 2^{-s} + 4^{-s} - 5^{-s} + 27^{-s} - 28^{-s} + \ldots,$$

in which $a_n = 1$ if $n = m^m$, $a_n = -1$ if $n = m^m + 1$, and $a_n = 0$ otherwise, we have

$$\bar{\sigma} = \sigma_0 = 0, \quad \sigma_\kappa = -\kappa.*$$

(6) The series
$$\Sigma \frac{(-1)^n n^a}{(\log n)^s} \qquad (a > 0)$$

is summable (λ, κ), where $\kappa > a$, for all values of s, but is never summable (λ, κ) if $\kappa < a$. It is summable (λ, a) if $\sigma \geqq -a$, and summable (l, a) if $\sigma > -a$.

(7) The series
$$\Sigma \frac{n^{-1-ai} (\log n)^\beta}{(\log \log n)^s} \qquad (a \neq 0, \ \beta > 0)$$

is summable everywhere by typical means of the first or second kind (or indeed by ordinary logarithmic means) of order greater than β, but is never summable by arithmetic means of any order.

* Bohr, 2.

VII

FURTHER DEVELOPMENTS OF THE THEORY OF FUNCTIONS REPRESENTED BY DIRICHLET'S SERIES

1. We shall now use the idea of summation by typical means to obtain generalisations of some of the most important theorems of Section III.

THEOREM 38. *Suppose the series* $\Sigma a_n e^{-\lambda_n s}$ *summable or finite* (λ, κ) *for* $s=\beta$. *Then*

$$f(s) = o(|t|^{\kappa+1})$$

uniformly for $\sigma \geqq \beta + \epsilon > \beta$.

We may plainly suppose, without loss of generality, that $\beta = 0$. There is a constant M such that

$$|A^\kappa(\tau)| < M\tau^\kappa$$

for all positive values of τ. Now, by Theorem 24,

$$f(s) = s^{\kappa+1}\int_{\lambda_1}^{\infty} A^\kappa(\tau)\, e^{-s\tau} d\tau$$

for $\sigma > 0$. Hence

$$|f(s)| < M|s|^{\kappa+1}\int_{\lambda_1}^{\infty} \tau^\kappa e^{-\sigma\tau}\, d\tau = M\left(\frac{|s|}{\sigma}\right)^{\kappa+1}\int_0^{\infty} v^\kappa e^{-v}\, dv$$

$$\leqq M(\sec a)^{\kappa+1}\,\Gamma(\kappa+1) = O(1) = o(|t|^{\kappa+1}),$$

uniformly in any angle of the type **a** of Lemma 9. Hence, in proving the theorem, we may confine ourselves to the parts of the half-plane $\sigma \geqq \epsilon$ which lie outside this angle; and so we may suppose $|s/t|$ less than a constant cosec a. This being so, we have

$$f(s) = s^{\kappa+1}\left(\int_{\lambda_1}^{\nu} + \int_{\nu}^{\infty}\right) A^\kappa(\tau)\, e^{-s\tau}\, d\tau = J_1 + J_2,$$

say.

Now $$|J_2| < M|s|^{\kappa+1}\int_{\nu}^{\infty} \tau^\kappa e^{-\sigma\tau}\, d\tau$$

$$\leqslant M(\operatorname{cosec} a)^{\kappa+1}|t|^{\kappa+1}\int_{\nu}^{\infty} \tau^\kappa e^{-\epsilon\tau}\, d\tau;$$

and so if δ is any positive number, we have $|J_2| < \delta|t|^{\kappa+1}$ for all values of ν greater than a number ν_0 which depends on ϵ and δ but not on s. When ν_0 has been chosen so that this inequality may hold, we have, by integration by parts,

$$J_1 = -s^\kappa e^{-s\nu} A^\kappa(\nu) + s^\kappa \int_{\lambda_1}^{\nu} e^{-s\tau} \frac{dA^\kappa(\tau)}{d\tau} d\tau\,* ;$$

and so, since $|e^{-s\tau}| < 1$,

$$|J_1| < H(\nu)|s|^\kappa \leqq H(\nu)(\operatorname{cosec}\alpha)^\kappa |t|^\kappa,$$

where $H(\nu)$ depends on ν alone. Accordingly

$$|f(s)| < H(\nu)(\operatorname{cosec}\alpha)^\kappa |t|^\kappa + \delta|t|^{\kappa+1} < 2\delta|t|^{\kappa+1},$$

if $|t|$ is large enough. Thus the theorem is proved†.

2. Generalisation of Theorem 13. We proceed next to a generalisation of Perron's formula discussed in § 2 of Section III.

THEOREM 39. *If the series is summable* (λ, κ), *where* $\kappa > 0$, *for* $s = \beta$, *and* $c > 0$, $c > \beta$, *then*

$$\frac{1}{\Gamma(\kappa+1)} \sum_{\lambda_n < \omega} a_n(\omega - \lambda_n)^\kappa = \frac{1}{2\pi i}\int_{c-i\infty}^{c+i\infty} \frac{f(s)}{s^{\kappa+1}} e^{\omega s} ds \quad \ldots\ldots(1).$$

This theorem depends on a generalisation of Lemma 3.

LEMMA 10. *We have*

$$\frac{1}{2\pi i}\int_{c-i\infty}^{c+i\infty} \frac{e^{us}}{s^{\kappa+1}} ds \begin{cases} = \dfrac{u^\kappa}{\Gamma(\kappa+1)} & (u \geqq 0), \\ = 0 & (u \leqq 0), \end{cases}$$

c and κ *being positive.*

We leave the verification of this lemma to the reader. It may be deduced without difficulty, by means of Cauchy's Theorem, from Hankel's expression of the reciprocal of the Gamma-function as a contour integral‡.

Let us suppose§ that $\lambda_m < \omega < \lambda_{m+1}$, and write

$$g(s) = e^{\omega s}\left\{f(s) - \sum_1^m a_n e^{-\lambda_n s}\right\} = \sum_{m+1}^{\infty} a_n e^{-(\lambda_n - \omega)s}.$$

Then what we have to prove reduces, in virtue of the lemma, to showing that

$$\int_{c-i\infty}^{c+i\infty} \frac{g(s)}{s^{\kappa+1}} ds = 0.$$

* The subject of integration may (if $\kappa < 1$) have isolated infinities across which it is absolutely integrable, but the integration by parts is permissible in any case.

† The result of the theorem is true, *a fortiori*, if (l, κ) be substituted for (λ, κ). It was given in this form, for integral values of κ, and for ordinary Dirichlet's series, by Riesz, **1**, and Bohr, **2**, **5**.

‡ Hankel, **1**; see also Heine, **1**, and Whittaker and Watson, *Modern Analysis* (ed. 2), p. 238. For a proof of results equivalent to those of Lemma 10, without the use of Cauchy's Theorem, see Dirichlet's *Vorlesungen über die Lehre von den einfachen und mehrfachen bestimmten Integralen* (ed. Arendt), pp. 166 *et seq.* The formulae may, in substance, be traced back to Cauchy.

§ Cf. III, § 2.

We have $$g(s) = e^{-(\lambda_{m+1} - \omega)s} h(s),$$
where $$h(s) = a_{m+1} + a_{m+2} e^{-(\lambda_{m+2} - \lambda_{m+1})s} + \dots.$$
This series is summable (μ, κ), where $\mu_n = \lambda_{m+n} - \lambda_{m+1}$, for $s = \beta$.* Hence $h(s) = o(|t|^{\kappa+1})$, uniformly for $\sigma \geqq c$. This relation replaces the equation $h(s) = o(|t|)$ used in the proof of Theorem 13; and the proof of Theorem 39 now follows exactly the same lines as that of the latter theorem. The final formula is valid even when $\omega = \lambda_n$, as the left-hand side is a continuous function of ω, and the integral is uniformly convergent.

More generally we have

$$\frac{1}{\Gamma(\kappa+1)} \sum_{\lambda_n < \omega} a_n e^{-\lambda_n s_0} (\omega - \lambda_n)^\kappa = \frac{1}{2\pi i} \int_{c-i\infty}^{c+i\infty} \frac{f(s)}{(s-s_0)^{\kappa+1}} e^{\omega(s-s_0)} \, ds \quad \dots(2),$$

if $c > \sigma_0$, $c > \beta$.

It is important for later applications to observe that the range of validity of the formulae (1) and (2) may be considerably extended. Let us suppose only that the series is summable (λ, κ) for *some* values of s, say for $\sigma > d$, and that the function $f(s)$ thus defined is regular for $\sigma > \beta$, where $\beta < d$, and satisfies the equation

$$f(s) = o(|t|^{\kappa+1}) \quad \dots\dots\dots\dots\dots\dots\dots\dots(3)$$

uniformly for $\sigma \geqq \beta + \epsilon > \beta$, however small ϵ may be. Then the theorem tells us that

$$\frac{1}{\Gamma(\kappa+1)} \sum_{\lambda_n < \omega} a_n (\omega - \lambda_n)^\kappa = \frac{1}{2\pi i} \int_{\gamma-i\infty}^{\gamma+i\infty} \frac{f(s)}{s^{\kappa+1}} e^{\omega s} \, ds$$

if $\gamma > 0$, $\gamma > d$. But, applying Cauchy's Theorem to the rectangle formed by the points on the lines $\sigma = c$, $\sigma = \gamma$ whose ordinates are $-T_1$ and T_2, and observing that, in virtue of (3), the contributions of the sides of the rectangle parallel to the real axis tend to zero when T_1 and T_2 tend to infinity, we see that the equation (1) still holds. A similar extension may be given to (2).

3. Analogous formulae for means of the second kind. There is a companion theorem to Theorem 39, viz.

THEOREM 40. *If* $\sum a_n e^{-\lambda_n s} = \sum a_n l_n^{-s}$ *is summable* (l, κ) *for* $s = \beta$, *and* $c > 0$, $c > \beta$, *then*

$$w^{-\kappa} \sum_{l_n < w} a_n (w - l_n)^\kappa = \frac{1}{2\pi i} \int_{c-i\infty}^{c+i\infty} f(s) \frac{\Gamma(\kappa+1)\Gamma(s)}{\Gamma(\kappa+1+s)} w^s \, ds \quad \dots\dots(1).\dagger$$

* See IV, § 4, (9), (10).

† The quotient of Γ-functions which figures under the sign of integration reduces, when κ is integral, to the κ-th difference of $1/s$.

As Theorem 39 depends on Lemma 10, so Theorem 40 depends upon

LEMMA 11. *If* $c>0$ *then*

$$\frac{1}{2\pi i}\int_{c-i\infty}^{c+i\infty}\frac{\Gamma(\kappa+1)\Gamma(s)}{\Gamma(\kappa+1+s)}v^s\,ds=\left(1-\frac{1}{v}\right)^\kappa \quad (v\geqq 1),$$
$$=0 \quad (v\leqq 1).$$

If we write
$$(1-x)^\kappa=\sum_0^\infty B_r^\kappa x^r$$

we have
$$\frac{\Gamma(\kappa+1)\Gamma(s)}{\Gamma(\kappa+1+s)}=\int_0^1 x^{s-1}(1-x)^\kappa\,dx=\sum_0^\infty\frac{B_r^\kappa}{s+r} \quad \text{..............(2).}$$

If we observe that

$$\frac{1}{2\pi i}\int_{c-i\infty}^{c+i\infty}\frac{v^s}{s+r}ds=v^{-r}\ (v>1),\quad =\frac{1}{2}\ (v=1)^*,\quad =0\ (v\leqq 1),$$

we see that the result of the lemma follows by substituting the series (2) under the sign of integration and integrating term by term. The details of the proofs of the lemma, and then of the theorem, present no particular difficulty, and we content ourselves with indicating the necessary formulae.

There is a generalisation of (1) corresponding to (2) of § 2, viz.

$$w^{-\kappa}\sum_{l_n<w}a_n l_n^{-s_0}(w-l_n)^\kappa=\frac{1}{2\pi i}\int_{c-i\infty}^{c+i\infty}f(s)\frac{\Gamma(\kappa+1)\Gamma(s-s_0)}{\Gamma(\kappa+1+s-s_0)}w^{s-s_0}ds\ \text{...(3)},$$

where $c>\sigma_0$, $c>\beta$. Both of the formulae (1) and (3), established originally on the hypothesis that $\sum a_n l_n^{-s}$ is summable (l, κ) for $s=\beta$, may then be extended to the case in which the series is only known to be summable (l, κ) for *some* values of s, and $f(s)$ satisfies the conditions stated at the end of § 2.

4. We are now in a position to consider an important group of theorems which differ fundamentally in character from those which we have considered hitherto. In such theorems as, for example, 23, 24, 27, 29, or 38, we start from the assumption that our series is summable for some particular value of s, and deduce properties of the function represented by the sum of the series. We shall now have to deal with theorems in which, to put the matter roughly, properties of the series are deduced from those of the function†.

One preliminary remark is necessary. When we speak of 'the function' we mean, of course, 'the function defined by means of, or associated with, the series.' That is to say, we imply that, for *some* values of s at any rate, *some* method of summation can be applied to

* If $v=1$ the principal value of the integral (in the sense explained in III, § 2) must be taken.

† The classical example of such a theorem is Taylor's Theorem, as proved by Cauchy for functions of a complex variable.

the series so as to give rise to the function. It is obviously, for our present purposes, the natural course to suppose that *for sufficiently large values of σ, say for $\sigma > d$, the series is summable by typical means of sufficiently high order.* There is thus an analytic function $f(s)$ associated with the series, and possibly capable of analytical continuation outside the known domain of summability of the series. In the theorems which follow we suppose that this is the case, and assume certain additional properties of $f(s)$. We then deduce from these properties more precise information as to the summability of the series.

5. THEOREM 41*. *Suppose that $f(s)$ is regular for $\sigma > \eta$, where $\eta < d$. Suppose further that κ and κ' are positive numbers such that $\kappa' < \kappa$, and that, however small be δ,*

$$f(s) = O(|s|^{\kappa'})$$

uniformly for $\sigma \geqq \eta + \delta > \eta$. Then $f(s)$ is summable (l, κ), and a fortiori summable (λ, κ), for $\sigma > \eta$.

If the series $\Sigma a_n l_n^{-s}$ is, for any values of s, summable (λ, κ), we know, by Theorem 40 and its extensions given at the end of § 3, that

$$w^{-\kappa} \sum_{l_n < w} a_n l_n^{-s_0} (w - l_n)^{\kappa} = \frac{1}{2\pi i} \int_{c - i\infty}^{c + i\infty} f(s) H(s - s_0) w^{s - s_0} ds \ldots (1),$$

where H is a certain product of Gamma-functions, provided only $c > \sigma_0$ and $c > \eta$. We can however free ourselves in this case from the assumption of the existence of a half-plane of summability (λ, κ). The series is summable (λ, k) somewhere, for *some* value of k, and therefore, if m is a sufficiently large positive integer, somewhere summable $(\lambda, \kappa + m)$. Hence we deduce the formula (1), with $\kappa + m$ in the place of κ. Now it will easily be verified that if we multiply (1) by w^{κ}, differentiate with respect to w, and divide by $\kappa w^{\kappa - 1}$, we obtain a formula which differs from (1) only in the substitution of $\kappa - 1$ for κ. Hence, by m differentiations, we can pass from $\kappa + m$ to κ. That the process of differentiation under the integral sign is legitimate follows at once from the relations

$$\frac{\Gamma(\kappa + p + 1)\Gamma(s - s_0)}{\Gamma(\kappa + p + 1 + s - s_0)} = O(|t|^{-\kappa - p - 1}), \quad f(s) = o(|t|^{\kappa'}),$$

where $\kappa' < \kappa$ and $p = 0, 1, \ldots, m - 1$; for the integrals obtained by differentiation are all absolutely and uniformly convergent.

* For $\lambda_n = \log n$, Riesz, **1**: in the general case, Riesz, **2**. Theorem 2 of the latter note includes Theorem 41 in virtue of Theorem 30.

Suppose now that $\eta < \sigma_0 < c$. Choose a number γ such that $\gamma > \eta$ and $\sigma_0 - 1 < \gamma < \sigma_0$, as is obviously possible. Then between the lines $\sigma = \gamma$, $\sigma = c$ lies one pole of $H(s - s_0)$, viz. $s = s_0$, with residue 1. Hence, by a simple application of Cauchy's Theorem, we obtain*

$$w^{-\kappa} \sum_{l_n < w} a_n l_n^{-s_0} (w - l_n)^\kappa - f(s_0) = \frac{1}{2\pi i} \int_{\gamma - i\infty}^{\gamma + i\infty} f(s)\, w^{s - s_0} H(s - s_0)\, ds \ldots (2).$$

But it is easy to see that the modulus of the integral is less than a constant multiple of $w^{\gamma - \sigma_0}$, and so tends to zero. Thus the theorem is established.

We add some remarks which will be of importance in the sequel. Let us suppose that $f(s)$ is *bounded* in every half-plane $\sigma \geqq \eta + \delta > \eta$. Then, if $\gamma = \sigma_0 - \theta$, where $0 < \theta < 1$, we have, for values of s situated on the line $\sigma = \gamma$,

$$|H(s - s_0)| < K |s - s_0|^{-\kappa - 1} = K\{(t - t_0)^2 + \theta^2\}^{-\frac{1}{2}(\kappa + 1)},$$

where K denotes a number which depends on κ and θ but not on σ_0 or t_0. Hence it follows that, throughout the domain $\sigma_0 \geqq \eta + \delta > \eta$, the integral on the right-hand side of (2) is less than a constant multiple of

$$w^{\gamma - \sigma_0} \int_{-\infty}^{\infty} \frac{dt}{\{(t - t_0)^2 + \theta^2\}^{\frac{1}{2}(\kappa + 1)}}.$$

This integral has obviously a value independent of t_0.† Hence it follows that *if $f(s)$ is limited in every half-plane $\sigma \geqq \eta + \delta > \eta$, the series is uniformly summable (l, κ) in every such half-plane, for any assigned positive value of κ.*

The same remarks apply as regards summability (λ, κ): they are not, as is the mere assertion of simple summability, immediate corollaries of the corresponding remarks concerning summability (l, κ); but it would be easy to complete Theorem 17 in such a way that they would become so. As we shall only make use of these remarks in the case of means of the second kind, it will not be necessary for us to go into this point in detail.

6. With Theorem 41 must be associated the following two more precise theorems.

* We apply Cauchy's Theorem to a rectangle whose shorter ends are made to tend to infinity. Since $f(s) = O(|s|^{\kappa'})$,

$$H(s - s_0) = O(|s|^{-\kappa - 1}),$$

and $\kappa' < \kappa$, the contributions of these ends tend to zero.

† It is important to observe that the argument would fail at this point if κ' were not zero.

THEOREM 42*. *If $f(s)$ is regular for $\sigma \geqq \eta$, except that it has, on the line $\sigma = \eta$, a finite number of poles or algebraical infinities of order less than $\kappa + 1$; if further*

$$f(s) = O(|s|^{\kappa'}),$$

where $0 \leqslant \kappa' < \kappa$, for $\sigma \geqq \eta$; then the series is uniformly summable (λ, κ) on any finite stretch of the line $\sigma = \eta$ which does not include any singular point.

THEOREM 43. *If the conditions of the preceding theorem are fulfilled, and the singularities on the line $\sigma = \eta$ are all algebraical infinities of order less than 1, then we may substitute (l, κ) for (λ, κ).*

We do not propose to insert proofs of these theorems†. We may add, however, that the results are capable of considerable generalisation. Thus the nature of the singularities permissible is considerably wider than appears from the enunciations. And in both theorems the hypothesis of regularity on the line $\sigma = \eta$ (except at a finite number of points) is quite unnecessarily restrictive. Thus in Theorem 42 this hypothesis might be replaced by that of continuity for $\sigma \geqq \eta$. In Theorem 43 this would not be sufficient; it would be necessary to impose restrictions similar to those which occur in the theory of the convergence of Fourier's series. The reader will find it instructive to consider the forms of the theorems when $\lambda_n = n$, remembering that summability (l, κ) is then equivalent to convergence (IV, § 4, (3)), and to compare them with the well-known theorems in the theory of Fourier's series to which they are then substantially equivalent.

The differences between Theorems 42 and 43 arise as follows. The formula (3) of § 3 represents the typical mean of the second kind *with its denominator* $w^{-\kappa}$, whereas the corresponding formula of § 2 represents that of the first kind without its denominator. Before studying the convergence of the latter mean the integral which occurs in (2) of § 2 must be divided by ω^{κ}; and it is owing to the presence of this factor that the means of the first kind converge under more general conditions. That the factor occurs in one case and not in the other is in its turn a consequence of the fact that the subject of integration has, for $s = s_0$, an infinity of order $\kappa + 1$ in the one case and order unity in the other.

There is another theorem which is also an interesting supplement to Theorem 41.

THEOREM 44. *If the series has a half-plane of absolute convergence, we can replace κ', in the enunciation of Theorem 41, by κ.*

We have $$f(s) = O(|t|^{\kappa})$$
for $\sigma = \eta + \delta$, and $f(s) = O(1)$ for $\sigma = \bar{\sigma} + \delta$, $\bar{\sigma}$ being the abscissa of absolute

* Riesz, **2**.

† The proofs depend on a combination of the arguments used in the proof of Theorem 41 with others similar to, but simpler than, those used by Riesz, **5** (pp. 98, 99), in proving and generalising Fatou's theorem (see VI, § 7, Theorem 37).

convergence, and δ any positive number. Hence by Lindelöf's Theorem (Theorem 14) we have

$$f(s)=O(|t|^{\kappa'}),$$

where $\kappa'=(\bar{\sigma}-\eta-\delta)\kappa/(\bar{\sigma}-\eta)<\kappa$, for $\sigma\geqq\eta+2\delta$. The result now follows from Theorem 41*.

7. From Theorems 38 and 41 we can deduce an important theorem first stated explicitly, for ordinary Dirichlet's series, by Bohr.

Since σ_κ is a decreasing function of κ, the numbers σ_κ tend to a limit, which may be $-\infty$, as $\kappa\rightarrow\infty$. We write

$$\lim_{\kappa\rightarrow\infty}\sigma_\kappa=S.$$

If S' is any number greater than S, the series is summable (λ, κ), for *some* value of κ, for $\sigma=S'$; and so, by Theorem 38, $f(s)$ is regular and of finite order (III, § 3) for $\sigma>S$. Conversely, if $f(s)$ is regular and of finite order for $\sigma>S'$, it follows from Theorem 41 that the series is summable (λ, κ), for sufficiently large values of κ, for $\sigma>S'$; and so $S'\geqq S$. Hence we deduce

THEOREM 45†. *If η is the least number such that $f(s)$ is regular and of finite order for $\sigma>\eta$, then $\eta=S$.*

8. The following two theorems are in a sense converses of Theorems 42 and 43.

THEOREM 46. *If $\Sigma a_n e^{-\lambda_n s}$ is summable (λ, κ) for $s=s_0$, then*

$$\lim_{\sigma\rightarrow\sigma_0}(\sigma-\sigma_0)^{\kappa+1}f(s)=0$$

uniformly throughout any finite interval of values of t.

THEOREM 47. *If the series is summable (l, κ) for $s=s_0$, we may replace $(\sigma-\sigma_0)^{\kappa+1}$ by $\sigma-\sigma_0$.*

The proofs of these theorems are simple. We indicate that of the first. We may obviously suppose, without loss of generality, that $s_0=0$ and $A=0$.

* This theorem includes a result given by Schnee, **7** (Theorems 3 and 3'). Schnee considers ordinary Dirichlet's series and Cesàro's means of integral order only. See the footnote (‖) to p. 23.

† This theorem was first enunciated in this form by Bohr, **2**. It is however, as shown above, an immediate consequence of Theorem 41 (or Theorem 3 of Riesz's note **1**). See also Bohr, **5**, **6**.

It follows from this theorem, for example, that if the Riemann hypothesis concerning the roots of the ζ-function is true, then the series $\Sigma\mu(n)n^{-s}$ is summable by arithmetic or logarithmic means for $\sigma>\frac{1}{2}$ (Bohr, **2**). As a matter of fact more than this is true: for it has been shown by Littlewood, **2**, that the Riemann hypothesis involves the convergence of the series for $\sigma>\frac{1}{2}$. The best previous result in this direction was due to Landau, **5**, and *H.*, p. 871.

We can then choose ν so that

$$|A^{\kappa}(\tau)| < \epsilon\tau^{\kappa} \qquad (\tau \geqq \nu).$$

By Theorem 24, we have

$$\sigma^{\kappa+1}|f(s)| < \frac{\sigma^{\kappa+1}|s|^{\kappa+1}}{\Gamma(\kappa+1)} \int_0^{\infty} |A^{\kappa}(\tau)| e^{-\sigma\tau} d\tau,$$

and $|s|$ is less than a constant throughout the region under consideration. Hence the preceding expression is less than a constant multiple of

$$\frac{\sigma^{\kappa+1}}{\Gamma(\kappa+1)} \int_0^{\nu} |A^{\kappa}(\tau)| d\tau + \frac{\epsilon\sigma^{\kappa+1}}{\Gamma(\kappa+1)} \int_0^{\infty} \tau^{\kappa} e^{-\sigma\tau} d\tau = \sigma^{\kappa+1} M(\nu) + \epsilon,$$

where $M(\nu)$ depends only on ν; and so is less than 2ϵ when σ is small enough. This proves the theorem: the proof of Theorem 47 is similar, starting from the integral representation of Theorem 29.

From Theorem 47 it follows that the series Σn^{-s} cannot be summable by any arithmetic mean on the line $\sigma = 1$, since the function $\zeta(s)$ has a pole of order 1 at $s = 1$.* On the other hand it follows from Theorem 42, and from the fact that $\zeta(1+ti) = O(\log|t|)$,† that it is summable by any logarithmic mean of positive order at all points of the line save $s = 1$.‡ Compare IV, § 4 (6).

9. Some theorems concerning ordinary Dirichlet's series. All the theorems of this section have been theorems concerning the most general type of Dirichlet's series. We pass now to a few theorems of a more special character. These theorems are valid for forms of λ_n whose rate of increase is sufficiently regular and not too much slower than that of $\log n$: we shall be content to prove them in the simplest and most interesting case, that in which $\lambda_n = \log n$.

THEOREM 48§. *If* $\Sigma a_n n^{-s}$ *is summable* (n, κ)‖ *for* $s = s_0$, *it is uniformly summable* (n, κ'), *where* κ' *is the greater of the numbers* $\kappa - \beta$ *and* 0, *in the domain*

$$\sigma \geqq \sigma_0 + \beta, \quad |t| \leqslant T.¶$$

The proof of this theorem is very similar to that of Theorems 23 and 29. We shall consider the case in which $0 < \kappa < 1$. We suppose, as we may do without loss of generality, that $s_0 = 0$ and $A = 0$. We choose a value of β such that $\kappa - \beta \geqq 0$, and we consider the arithmetic

* Landau, *H.*, p. 161.

† Landau, *H.*, p. 169.

‡ For further results relating to the series for $\{\zeta(s)\}^c$ see Riesz, **1**, **2**.

§ For integral orders of summation, and $\sigma > \sigma_0 + \beta$, Bohr, **1**; in the general form, Riesz, **1**.

‖ *I.e.* by *arithmetic* means of order κ.

¶ These inequalities might be replaced by $\sigma \geqq \sigma_0 + \beta$, $|\mathrm{am}\, s| \leqq \alpha < \frac{1}{2}\pi$.

mean of order $\kappa - \beta$ at a point s for which $\sigma \geqq \beta$. This mean is easily seen to be (cf. VI, §§ 2 and 5)

$$-w^{-\kappa+\beta}\int_1^w A(u)\frac{d}{du}\{u^{-s}(w-u)^{\kappa-\beta}\}\,du$$

$$=-w^{-\kappa+\beta}\int_1^w A(u)\frac{d}{du}\{(u^{-s}-w^{-s})(w-u)^{\kappa-\beta}\}\,du$$

$$-(\kappa-\beta)\,w^{-\kappa+\beta-s}\int_1^w A(u)(w-u)^{\kappa-\beta-1}\,du.$$

The second term is

$$-w^{-\kappa+\beta-s}A^{\kappa-\beta}(w),$$

and, by Theorem 22, $A^{\kappa-\beta}(w) = o(w^{\kappa})$. Hence this term is of the form $o(1)$, uniformly for $\sigma \geqq \beta$. The first term we integrate by parts, obtaining

$$w^{-\kappa+\beta}\int_1^w A^1(u)\frac{d^2}{du^2}\{(u^{-s}-w^{-s})(w-u)^{\kappa-\beta}\}\,du = J_1 + J_2 + J_3,$$

say, where J_1, J_2, and J_3 contain under the sign of integration respectively factors

$$s(s+1)\,u^{-s-2}(w-u)^{\kappa-\beta},\quad 2s(\kappa-\beta)\,u^{-s-1}(w-u)^{\kappa-\beta-1},$$

$$(\kappa-\beta)(\kappa-\beta-1)(u^{-s}-w^{-s})(w-u)^{\kappa-\beta-2}.$$

We can now show, by arguments resembling those of VI, § 3 so closely that it is hardly necessary to set them out at length, that J_1 tends to the limit

$$s(s+1)\int_1^\infty A^1(u)\,u^{-s-2}\,du,$$

and J_2 and J_3 to zero, uniformly in the region $\sigma \geqq \beta$, $|t| \leqq T$.

THEOREM 49. *If the series is summable* (n, κ), *uniformly for* $\sigma = \sigma_0$, *it is summable* (n, κ'), *uniformly for* $\sigma \geqq \sigma_0 + \beta$.

We apply the argument used in the proof of Theorem 48 to pass from the point $\sigma_0 + it$ to the point $\sigma + it$ with the same ordinate, and take account of the uniformity postulated on the line $\sigma = \sigma_0$. The result then follows substantially as before.

10. By combining Theorems 41 and 48 we arrive at the following theorem.

THEOREM 50. *If, however small* δ *and* ϵ *may be, we have*

$$f(s) = O(|t|^{\epsilon})$$

in the half-plane $\sigma \geqq \eta + \delta$, *then the series is convergent in every such half-plane, i.e. for* $\sigma > \eta$.

For, by Theorem 41, the series is summable (n, ϵ_1), where ϵ_1 is any number greater than ϵ, for $\sigma > \eta + \delta$. Hence, by Theorem 48, it is convergent for $\sigma > \eta + \delta + \epsilon_1$, *i.e.* for $\sigma > \eta$.

Theorem 50 is but a particular case of an important theorem generally known as the 'Schnee-Landau' Theorem.

THEOREM 51. *If* $a_n = O(n^\delta)$ *for all positive values of* δ, *so that the series is absolutely convergent for* $\sigma > 1$, *and*

$$f(s) = O(|t|^k) \qquad (k > 0)$$

uniformly for $\sigma > \eta$, *then the series is convergent for* $\sigma > \zeta$, *where* ζ *is the lesser of the numbers*

$$\frac{\eta + k}{1 + k}, \quad \eta + k.^*$$

To deal with this theorem and its generalisations would require more space than is at our disposal here, and we must be content to refer to Landau's *Handbuch* and to the original memoirs by Landau and Schnee†. If the second condition is satisfied for all positive values of k, then the series is convergent for $\sigma > \eta$. The first condition then becomes unnecessary, as may be seen at once by applying a linear transformation to the variable s; and so we obtain Theorem 50.

11. We can obtain another important theorem by combining Theorem 49 with the result proved at the end of § 5. Suppose that $f(s)$ is bounded in every half-plane $\sigma \geqq \eta + \delta > \eta$. Then, if δ and ϵ are chosen arbitrarily, the series is uniformly summable (n, ϵ) for $\sigma \geqq \eta + \delta$, and therefore, by Theorem 49, uniformly convergent for $\sigma \geqq \eta + \delta + \epsilon$. We thus obtain

THEOREM 52. *If* $f(s)$ *is bounded in every half-plane* $\sigma \geqq \eta + \delta > \eta$, *then the series is uniformly convergent in every such half-plane.*

This theorem was first given by Bohr‡. Its converse is obviously trivial.

Before leaving these theorems we may make a few additional remarks. Theorems 48—52 may be extended§ to any type of series $\Sigma a_n e^{-\lambda_n s} = \Sigma a_n l_n^{-s}$ for which a positive constant g exists such that

$$\frac{l_{n+1}}{l_{n+1} - l_n} = O(l_n^g) \qquad (1).$$

* The first number gives the better result if $\eta + k > 0$, the second if $\eta + k < 0$.

† Landau, *H.*, pp. 853 *et seq.* See also Landau, **5**, **7**; Schnee, **4**, **7**; Bohr, **2**, **5**, **10**.

‡ Bohr, **3**, **8**.

§ See the memoirs cited in footnote (†).

This hypothesis ensures that the increase of l_n is *not too slow*; it is satisfied, for instance, if $l_n=n$ or $l_n=e^n$, but not if $l_n=\log n$. It is easy to show that the condition (1) is equivalent to either of the following:

$$l_n^{-h}=O(l_{n+1}-l_n) \quad \text{.............................(2)},$$

$$e^{-k\lambda_n}=O(\lambda_{n+1}-\lambda_n) \quad \text{.............................(3)},$$

where h and k are positive*.

Considerations of space forbid us from giving details of these generalisations. We would only warn the reader that the proofs, involving as they do in some places an appeal to the delicate Theorem 22, are not entirely simple, especially when the increase of l_n is very rapid and irregular.

The line $\sigma=\eta$ such that the series is uniformly convergent for $\sigma \geqq \eta+\delta$, but not for $\sigma \geqq \eta-\delta$, however small be δ, has been called by Bohr the *line of uniform convergence*. It has been shown by Bohr† that, when the numbers λ_n are linearly independent, the line of uniform convergence is identical with the line of absolute convergence: but he has given an example of a series (naturally corresponding to a non-independent sequence of λ's) which possesses a half-plane of uniform convergence and no half-plane of absolute convergence.

12. Convexity of the abscissa σ_κ, considered as a function of κ. It was shown by Bohr ‡ that the abscissa of summability σ_r, of integral order, belonging to an ordinary Dirichlet's series, satisfies the inequalities

$$\sigma_{r+1} \leqq \sigma_r \leqq \sigma_{r+1}+1 \quad \text{........................(1)},$$

$$\sigma_r-\sigma_{r+1} \geqq \sigma_{r+1}-\sigma_{r+2} \quad \text{....................(2)}.$$

Of the inequalities (1), the first is an obvious corollary of Theorem 16 (cf. VI, § 4); and the second is an obvious corollary of Theorem 48. The inequalities (2) lie deeper.

The property which is expressed by the inequalities (2) was then considered by Hardy and Littlewood §, who proved more precise theorems of which Bohr's inequalities are corollaries. Their results have since been extended by Riesz ||, so as to apply to the most general type of Dirichlet's series and to all orders of summation integral or non-integral. In particular it has been proved that *the abscissa σ_κ is in all cases a convex function of κ.*

It was also shown by Bohr¶ that the conditions (1) and (2) are

* These conditions are rather wider than that adopted by Schnee and Landau, and are substantially the same as that adopted by Bohr. It is natural to suppose h and k positive, but not necessary; for if (3), for example, is satisfied with $k \leqq 0$, it is plainly satisfied with $k>0$.

† Bohr, **7**: cf. III, § 7. ‡ Bohr, **2**, **5**. § Hardy and Littlewood, **2**.

|| In a memoir as yet unpublished. ¶ Bohr, **5**.

necessary and sufficient that a given sequence σ_r should be the abscissae of summability of *some* ordinary Dirichlet's series*.

13. Summation of Dirichlet's Series by other methods. It is natural to enquire whether methods of summation different in principle from those which we have considered may not be useful in the theory. The first to suggest itself is Borel's exponential method. The application of this method to ordinary Dirichlet's series has been considered by Hardy and by Fekete†. It has been shown, for example, that the regions of summability, and of absolute summability, are half-planes; and that the method at once gives the analytical continuation all over the plane of certain interesting classes of series. But the method is not one which seems likely to render great services to the general theory.

Riesz‡ has considered methods of summation related to Borel's, and its generalisation by Mittag-Leffler, somewhat as the typical means of this section are related to Cesàro's original means. These methods lead to representations of the function associated with the series which differ fundamentally in one very important respect from those afforded by the theory of typical means. Their domains of application may, like Borel's polygon of summability, or Mittag-Leffler's *étoile*, be defined simply by means of the singular points of the function, and necessarily contain singular points on their frontier.

VIII

THE MULTIPLICATION OF DIRICHLET'S SERIES

1. We shall be occupied in this section with the study of a special problem, interesting on account of the variety and elegance of the results to which it has led, and important on account of its applications in the Analytic Theory of Numbers§.

* The construction given by Bohr (*l.c.* pp. 127 *et seq.*), for a series with given abscissae may be simplified by using the series $\Sigma e^{Ain^a} n^{-s}$ of IV, § 4, (5) as a 'simple element' in place of the series which he uses.

† Hardy, **3**; Fekete, **1**.

‡ Riesz, **6**.

§ In this connection we refer particularly to Landau, **4**, and *H.*, pp. 750 *et seq.*

We denote by A and B the series

$$a_1 + a_2 + \dots, \quad b_1 + b_2 + \dots,$$

and by C the 'product-series'

$$c_1 + c_2 + \dots,$$

where c_n is a function of the a's and b's, to be defined more precisely in a moment. We shall also use A, B, C to denote the *sums* of the series, when they are convergent or summable.

When C is formed in accordance with Cauchy's rule*, we have

$$c_p = a_1 b_p + a_2 b_{p-1} + \dots + a_p b_1 = \sum_{m+n=p+1} a_m b_n.$$

Cauchy's rule for multiplication is, however, only one among an infinity. We are led to it by arranging the formal product of the power series $\sum a_m x^m$, $\sum b_n x^n$ in powers of x and putting $x = 1$, or, what is the same thing, by arranging the formal product of the Dirichlet's series

$$\sum a_m e^{-ms}, \quad \sum b_n e^{-ns}$$

according to the ascending order of the sums $m + n$, associating together all the terms for which $m + n$ has the same value, and then putting $s = 0$. It is clear that we arrive at a generalisation of our conception of multiplication by considering the general Dirichlet's series

$$\sum a_m e^{-\lambda_m s}, \quad \sum b_n e^{-\mu_n s}$$

and arranging their formal product according to the ascending order of the sums $\lambda_m + \mu_n$. Let (ν_p) be the ascending sequence formed by all the values of $\lambda_m + \mu_n$†. Then the series $C = \sum c_p$, where

$$c_p = \sum_{\lambda_m + \mu_n = \nu_p} a_m b_n,$$

will be called *the Dirichlet's product of the series A, B, of type* (λ, μ).

Thus if $\lambda_m = \log m$, $\mu_n = \log n$, so that we are dealing with ordinary Dirichlet's series, then $\nu_p = \log p$ and

$$c_p = \sum_{mn=p} a_m b_n = \sum_d a_d b_{p/d},$$

the latter summation extending to all the divisors d of p.

* See *e.g.* Bromwich, *Infinite series*, p. 83.

† It is generally the case in applications that the λ and μ sequences are the same. Any case can be formally reduced to this case by regarding all the numbers λ_m and μ_n as forming one sequence and attributing to each series a number of terms with zero coefficients (Landau, *H.*, p. 750). In the most important cases (*e.g.* $\lambda_m = m$, $\lambda_m = \log m$) the ν sequence is also the same, but of course this is not generally true. In the theoretically general case no two values of $\lambda_m + \mu_n$ will be equal.

2. The three classical theorems relating to ordinary multiplication (Cauchy's, Mertens', and Abel's) have their analogues in the general theory.

THEOREM 53. *If A and B are absolutely convergent, then C is absolutely convergent and $AB = C$.*

This theorem is merely a special case of the classical theorem which asserts that the absolutely convergent double series $\Sigma\, a_m b_n$ may be summed indifferently in any manner we please*.

THEOREM 54. *If A is absolutely convergent and B convergent, then C is convergent and $AB = C$.*†

We shall prove that $\Sigma\, a_m b_n$ converges to the sum AB when arranged as a simple series so that $a_m b_n$ comes before $a_{m'} b_{n'}$ if $\lambda_m + \mu_n < \lambda_{m'} + \mu_{n'}$ (the order of the terms for which $\lambda_m + \mu_n$ has the same value being indifferent). Theorem 54 then follows by bracketing all the terms for which $\lambda_m + \mu_n$ has the same value.

Suppose first that $B = 0$. Let S_ν be any partial sum of the new series, and let a_k be the a of highest rank that occurs in it. Then

$$S_\nu = \sum_{p=1}^{k} a_p B_r,$$

where r is a function of k and p. Suppose that S_ν contains a term $a_\gamma b_\gamma$. Then it contains *all* the terms

$$a_p b_q \qquad (p \leqq \gamma,\ q \leqq \gamma).\ddagger$$

Thus $k \geqq \gamma$ and $r \geqq \gamma$ for $p = 1, 2, \dots \gamma$.

Now we can choose γ so that

$$|B_r| < \epsilon \qquad (r \geqq \gamma),$$

and

$$\sum_{\gamma+1}^{\infty} |a_p| < \epsilon.$$

Then

$$|S_\nu| < \epsilon \sum_{1}^{\gamma} |a_p| + M \sum_{\gamma+1}^{k} |a_p| < \epsilon(\mathbf{A} + M),$$

where $\mathbf{A}$ denotes the sum of the series $\Sigma\, |a_p|$ and M is any number greater than the greatest value of $|B_r|$. Thus $S_\nu \to 0$ as $\gamma \to \infty$, that is to say as $k \to \infty$.

* See *e.g.* Bromwich, *Infinite series*, p. 81. This theorem is not merely a special case of Theorem 54, because it asserts the *absolute* convergence of the product series.

† Stieltjes, **2**; Landau, **4**, and *H.*, p. 752. See also Wigert, **1**.

‡ This is the kernel of the proof. The reader will find that a figure will help to elucidate the argument.

Secondly, suppose $B \neq 0$. We form a new series B' for which

$$b_1' = b_1 - B, \quad b_2' = b_2, \quad b_3' = b_3, \ldots.$$

Then, by what precedes, $\sum a_m b_n'$ converges to zero, and so $\sum a_m b_n$ converges to AB.

3. THEOREM 55. *If the series A, B, C are all convergent, then $AB = C$.*

This is the analogue of Abel's theorem for power series*. We shall deduce it from a more general theorem, the analogue for Dirichlet's series of a well-known theorem of Cesàro†.

THEOREM 56. *If A is summable (λ, α) and B is summable (μ, β), then C is summable $(\nu, \alpha + \beta + 1)$, and $AB = C$.*

If $\gamma = \alpha + \beta + 1$ we have

$$A_\lambda^\alpha(\omega) = \sum a_m (\omega - \lambda_m)^\alpha, \quad B_\mu^\beta(\omega) = \sum b_n (\omega - \mu_n)^\beta, \quad C_\nu^\gamma(\omega) = \sum c_p (\omega - \nu_p)^\gamma,$$

the summations being limited respectively by the inequalities $\lambda_m < \omega$, $\mu_n < \omega$, $\nu_p < \omega$. Then

$$C_\nu^\gamma(\omega) = \frac{\Gamma(\gamma+1)}{\Gamma(\alpha+1)\Gamma(\beta+1)} \int_0^\omega A_\lambda^\alpha(\tau) B_\mu^\beta(\omega - \tau)\, d\tau \quad \ldots\ldots(1).$$

For consider the term $a_m b_n$. It occurs in $C_\nu^\gamma(\omega)$ if $\lambda_m + \mu_n < \omega$, and its coefficient is

$$(\omega - \lambda_m - \mu_n)^\gamma.$$

The term a_m occurs in $A_\lambda^\alpha(\tau)$ if $\lambda_m < \tau$, with coefficient $(\tau - \lambda_m)^\alpha$, and b_n occurs in $B_\mu^\beta(\omega - \tau)$ if $\mu_n < \omega - \tau$, with coefficient $(\omega - \tau - \mu_n)^\beta$. Hence $a_m b_n$ occurs on the right-hand side of (1) if $\lambda_m + \mu_n < \omega$, and its coefficient is

$$\frac{\Gamma(\gamma+1)}{\Gamma(\alpha+1)\Gamma(\beta+1)} \int_{\lambda_m}^{\omega-\mu_n} (\tau - \lambda_m)^\alpha (\omega - \tau - \mu_n)^\beta\, d\tau = (\omega - \lambda_m - \mu_n)^\gamma.$$

* The theorem was first given by Landau, **4**, in the case in which at least one of the series $\sum a_m e^{-\lambda_m s}$, $\sum b_n e^{-\mu_n s}$ possesses a region of absolute convergence. His proof depended on considerations of function-theory. A purely arithmetic and completely general proof was then discovered independently by Phragmén, Riesz, and Bohr. This proof depends on the particular case of Theorem 56 in which $\alpha = \beta = 0$. See Landau, *H.*, pp. 762, 904; Riesz, **2**; Bohr, **2**.

† Cesàro, **1**; see also Bromwich, *Infinite series*, p. 316. Cesàro and Bromwich consider only integral orders of summability. The extension to non-integral orders is due to Knopp, **2**, and Chapman, **1**.

Thus (1) is established. But

$$A_{\lambda}^{\alpha}(\tau) \sim A\tau^{\alpha}, \quad B_{\mu}^{\beta}(\tau) \sim B\tau^{\beta};$$

and therefore, by Lemma 5,

$$C_{\nu}^{\gamma}(\omega) \sim AB\,\omega^{\gamma}.$$

This proves the theorem. In particular, if A and B are convergent, the product series is summable $(\nu, 1)$. Theorem 55 then follows from Theorems 56 and 16.

4. The following generalisation of Theorem 54 provides an interesting companion theorem to Theorem 56.

THEOREM 57. *If A is absolutely convergent, and B is summable (μ, β), then C is summable (ν, β) and $C = AB$.**

In this theorem the λ-sequence is at our disposal. It is evidently enough (cf. § 2) to prove the theorem in the particular case when $B = 0$.

We have

$$C_{\nu}^{\beta}(\omega) = \sum_{\lambda_m + \mu_n < \omega} a_m b_n (\omega - \lambda_m - \mu_n)^{\beta} = \sum_{\lambda_m < \omega} a_m B_{\mu}^{\beta}(\omega - \lambda_m).$$

There is a constant M such that

$$|B_{\mu}^{\beta}(\tau)| < M\tau^{\beta}$$

for all values of τ; and we can choose ω so that (1) the M on the right-hand side of this inequality can be replaced by ϵ if $\tau_m \geqq \frac{1}{2}\omega$, and (2)

$$\sum_{\frac{1}{2}\omega \leqq \lambda_m < \omega} |a_m| < \epsilon.$$

Then we have

$$|C_{\nu}^{\beta}(\omega)| < M \sum_{\frac{1}{2}\omega \leqq \lambda_m < \omega} |a_m| (\omega - \lambda_m)^{\beta} + \epsilon \sum_{\lambda_m < \omega} |a_m| (\omega - \lambda_m)^{\beta},$$

$$|\omega^{-\beta} C_{\nu}^{\beta}(\omega)| < M \sum_{\frac{1}{2}\omega \leqq \lambda_m < \omega} |a_m| + \epsilon \mathbf{A} < \epsilon (M + \mathbf{A}),$$

and so

$$\omega^{-\beta} C_{\nu}^{\beta}(\omega) \to 0:$$

which proves the theorem.

5. The next theorem which we shall state is one whose general idea is analogous to those of the 'Tauberian' theorems of VI, § 7, and in particular Theorem 35. We therefore omit the proof.

* For the special case of multiplication in accordance with Cauchy's rule, see Hardy and Littlewood, **2** (Theorem 35), where further theorems on the multiplication of series will be found. The particular theorem proved there is however a special case of one given previously by Fekete, **3**.

THEOREM 58. *If*

$$a_m = O\left(\frac{\lambda_m - \lambda_{m-1}}{\lambda_m}\right), \quad b_n = O\left(\frac{\mu_n - \mu_{n-1}}{\mu_n}\right),$$

*then the convergence of A and B is enough to ensure that of C.**

6. Our last two theorems are of a different character.

THEOREM 59†. *If*

(1) $\tau \geqq 0,\ \tau' \geqq 0,\ \tau+\tau' > 0,\ \rho+\tau \geqq \rho',\ \rho'+\tau' \geqq \rho,$

(2) *the series $\Sigma a_m e^{-\lambda_m s}$ is convergent for $s=\rho$, and absolutely convergent for $s=\rho+\tau$,*

(3) *the series $\Sigma b_n e^{-\mu_n s}$ is convergent for $s=\rho'$, and absolutely convergent for $s=\rho'+\tau'$;*

then the series $\Sigma c_p e^{-\nu_p s}$ is convergent for

$$s = \frac{\rho\tau' + \rho'\tau + \tau\tau'}{\tau+\tau'}.$$

We shall give a proof of this theorem only in the simplest and most interesting case, viz. that in which

$$\lambda_m = \log m, \quad \mu_n = \log n, \quad \nu_p = \log p,$$

so that the series are ordinary Dirichlet's series, and

$$\rho = \rho' = 0.$$

We can then suppose that τ and τ' are any numbers greater than 1, so that

$$\frac{\rho\tau' + \rho'\tau + \tau\tau'}{\tau+\tau'}$$

may be any number greater than $\frac{1}{2}$. The theorem therefore asserts that, if A and B are convergent, and

$$c_p = \sum_{mn=p} a_m b_n,$$

then $\Sigma c_p p^{-s}$ is convergent for $\sigma > \frac{1}{2}$. In this case, however, it is possible to prove rather more.

* This theorem is not a corollary of Theorem 35. The conditions do not ensure that $c_p = O\{(\nu_p - \nu_{p-1})/\nu_p\}$. The theorem was proved, in the particular case $\lambda_m = m$, $\mu_n = n$, by Hardy, **2**; and in the general case by Hardy, **7**. Hardy however supposed the indices λ_m, μ_n subject to the conditions

$$\lambda_m - \lambda_{m-1} = o(\lambda_m), \quad \mu_n - \mu_{n-1} = o(\mu_n).$$

That these conditions are unnecessary was shown by Rosenblatt, **2**.

† Landau, **4**, and *H.*, p. 755.

THEOREM 60. *If A and B are convergent, then $\Sigma \frac{c_p}{\sqrt{p}}$ is convergent**.

We shall prove, in fact, that $\Sigma c_p p^{-s}$ is uniformly convergent along any finite stretch of the line $s=\frac{1}{2}+ti$.

Let us write
$$A_x = \sum_{m>x} a_m, \quad A = A(x) + A_x,$$
and similarly for B. We have
$$\sum_m^\infty a_\nu \nu^{-s} = \sum_m^\infty (A_{\nu-1} - A_\nu)\nu^{-s}$$
$$= A_{m-1} m^{-s} + \sum_m^\infty A_\nu \{\nu^{-s} - (\nu+1)^{-s}\} = o\left(\frac{1}{\sqrt{m}}\right),$$
as
$$m^{-s} = O\left(\frac{1}{\sqrt{m}}\right), \quad \sum_m^\infty |\nu^{-s} - (\nu+1)^{-s}| = O\left(\frac{1}{\sqrt{m}}\right), \quad A_m = o(1).$$

Similarly
$$\sum_n^\infty b_\nu \nu^{-s} = o\left(\frac{1}{\sqrt{n}}\right);$$
and these relations all hold uniformly as regards t. We observe now that
$$\sum_1^x c_p p^{-s}$$
includes all products of pairs of terms $a_m m^{-s}$, $b_n n^{-s}$ for which $mn \leqq [x]$, and
$$\sum_1^{\sqrt{x}} a_m m^{-s} \times \sum_1^{\sqrt{x}} b_n n^{-s}$$
all for which $m \leqq \sqrt{x}$, $n \leqq \sqrt{x}$; and that, if $mn \leqq [x]$, one at least of m and n is not greater than $\sqrt{x}$. It follows that
$$\sum_1^x c_p p^{-s} - \sum_1^{\sqrt{x}} a_m m^{-s} \sum_1^{\sqrt{x}} b_n n^{-s}$$
$$= \sum_1^{\sqrt{x}} a_m m^{-s} \sum_{\sqrt{x}}^{x/m} b_\nu \nu^{-s} + \sum_1^{\sqrt{x}} b_n n^{-s} \sum_{\sqrt{x}}^{x/n} a_\nu \nu^{-s}$$
$$= o(x^{-\frac{1}{4}}) \sum_1^{\sqrt{x}} \frac{1}{\sqrt{m}} + o(x^{-\frac{1}{4}}) \sum_1^{\sqrt{x}} \frac{1}{\sqrt{n}} = o(1);†$$
which proves the theorem.

It was suggested by Cahen‡ that the convergence of A and B should involve the convergence of $\Sigma c_p p^{-s}$ for $\sigma > 0$, and not merely for $\sigma \geqq \frac{1}{2}$ (as is shown by Theorem 60). This question, the answer to which remained for long doubtful, was ultimately decided by Landau§, who showed by an example that Cahen's hypothesis was untrue.

* Stieltjes, **1**, **2**. See also Landau, **4**, and *H.*, pp. 759 *et seq.*

† Since
$$\sum_{\sqrt{x}}^{x/m} b_\nu \nu^{-s} = o(x^{-\frac{1}{4}}) - o\left(\sqrt{\frac{m}{x}}\right) = o(x^{-\frac{1}{4}});$$
and similarly
$$\sum_{\sqrt{x}}^{x/n} a_\nu \nu^{-s} = o(x^{-\frac{1}{4}}).$$

‡ Cahen, **1**. § Landau, **5**, and *H.*, p. 773.

This may be seen very simply by means of Bohr's example (III, § 7) of a function $f(s)$, convergent for $\sigma>0$, for which $\mu(\sigma)=1-\sigma$ for $0<\sigma<1$. If we square this function, we obtain a function for which $\mu(\sigma)=2-2\sigma$ for $0<\sigma<1$, so that $\mu(\sigma)>1$ if $\sigma<\frac{1}{2}$. It follows from Theorem 12 that the squared series cannot converge for $\sigma<\frac{1}{2}$, and hence that the number $\frac{1}{2}$ which occurs in Theorem 60 cannot possibly be replaced by any smaller number*.

* Bohr, **5**.

BIBLIOGRAPHY

[The following list of memoirs does not profess to be exhaustive. It contains (1) memoirs actually referred to in the text, (2) memoirs which have appeared since the publication of Landau's *Handbuch* (in 1909) and which are concerned with the theory of summable series or the general theory of Dirichlet's series. We have added a few representative memoirs concerned with applications of the idea of summability in allied theories such as the theory of Fourier's series.]

N. H. Abel

(1) 'Untersuchungen über die Reihe $1+\frac{m}{1}x+\frac{m(m-1)}{1.2}x^2+\ldots$', *Journal für Math.*, vol. 1, 1826, pp. 311–339 (*Œuvres*, vol. 1, pp. 219–250).

F. R. Berwald

(1) 'Solution nouvelle d'un problème de Fourier', *Arkiv för Matematik*, vol. 9, 1913, no. 14, pp. 1–18.

H. Bohr

(1) 'Sur la série de Dirichlet', *Comptes Rendus*, 11 Jan. 1909.

(2) 'Über die Summabilität Dirichletscher Reihen', *Göttinger Nachrichten*, 1909, pp. 247–262.

(3) 'Sur la convergence des séries de Dirichlet', *Comptes Rendus*, 1 Aug. 1910.

(4) 'Beweis der Existenz Dirichletscher Reihen, die Nullstellen mit beliebig grosser Abszisse besitzen', *Rendiconti di Palermo*, vol. 31, 1910, pp. 235–243.

(5) 'Bidrag til de Dirichlet'ske Rækkers Theori', *Dissertation*, Copenhagen, 1910.

(6) 'Über die Summabilitätsgrenzgerade der Dirichletschen Reihen', *Wiener Sitzungsberichte*, vol. 119, 1910, pp. 1391–1397.

(7) 'Lösung des absoluten Konvergenzproblems einer allgemeinen Klasse Dirichletscher Reihen', *Acta Math.*, vol. 36, 1911, pp. 197–240.

(8) 'Über die gleichmässige Konvergenz Dirichletscher Reihen', *Journal für Math.*, vol. 143, 1913, pp. 203–211.

(9) 'Über die Bedeutung der Potenzreihen unendlich vieler Variabeln in der Theorie der Dirichletschen Reihen $\Sigma a_n n^{-s}$', *Göttinger Nachrichten*, 1913, pp. 441–488.

(10) 'Einige Bemerkungen über das Konvergenzproblem Dirichletscher Reihen', *Rendiconti di Palermo*, vol. 37, 1913, pp. 1–16.

(11) 'Darstellung der gleichmässigen Konvergenzabszisse einer Dirichletschen Reihe $\Sigma a_n n^{-s}$ als Funktion der Koeffizienten der Reihe', *Archiv der Math. und Phys.*, ser. 2, vol. 21, 1913, pp. 326–330.

H. Bohr and E. Landau

(1) 'Über das Verhalten von $\zeta(s)$ und $\zeta_\kappa(s)$ in der Nähe der Geraden $\sigma = 1$', *Göttinger Nachrichten*, 1910, pp. 303–330.

(2) 'Ein Satz über die Dirichletschen Reihen mit Anwendung auf die ζ-Funktion und die L-Funktionen', *Rendiconti di Palermo*, vol. 37, 1913, pp. 269–272.

T. J. I'A. Bromwich

(1) 'On the limits of certain infinite series and integrals', *Math. Ann.*, vol. 65, 1908, pp. 350–369.

(2) 'The relation between the convergence of series and of integrals', *Proc. Lond. Math. Soc.*, ser. 2, vol. 6, 1908, pp. 327–338.

E. Cahen

(1) 'Sur la fonction $\zeta(s)$ de Riemann et sur des fonctions analogues', *Ann. de l'École Normale*, ser. 3, vol. 11, 1894, pp. 75–164.

E. Cesàro

(1) 'Sur la multiplication des séries', *Bulletin des Sciences Math.*, vol. 14, 1890, pp. 114–120.

S. Chapman

(1) 'On non-integral orders of summability of series and integrals', *Proc. Lond. Math. Soc.*, ser. 2, vol. 9, 1910, pp. 369–409.

(2) 'On the general theory of summability, with applications to Fourier's and other series', *Quarterly Journal*, vol. 43, 1912, pp. 1–53.

(3) 'On the summability of series of Legendre's functions', *Math. Ann.*, vol. 72, 1912, pp. 211–227.

(4) 'Some theorems on the multiplication of series which are infinite in both directions', *Quarterly Journal*, vol. 44, 1913, pp. 219–233.

(*See also* **G. H. Hardy** and **S. Chapman**.)

G. Faber

(1) 'Über die Hölderschen und Cesàroschen Grenzwerte', *Münchener Sitzungsberichte*, vol. 43, 1913, pp. 519–531.

P. Fatou

(1) 'Séries trigonométriques et séries de Taylor', *Acta Math.*, vol. 30, 1906, pp. 335–400 (*Thèse*, Paris, 1907).

L. Fejér

(1) 'Untersuchungen über Fouriersche Reihen', *Math. Ann.*, vol. 58, 1904, pp. 51–69.

(2) 'Über die Laplacesche Reihe', *Math. Ann.*, vol. 67, 1909, pp. 76–109.

(3) 'La convergence sur son cercle de convergence d'une série de puissances effectuant une représentation conforme du cercle sur le plan simple', *Comptes Rendus*, 6 Jan. 1913.

(4) 'Über die Konvergenz der Potenzreihen auf der Konvergenzgrenze in Fällen der konformen Abbildung auf die schlichte Ebene', *H. A. Schwarz Festschrift*, 1914, pp. 42–53.

M. Fekete

(1) 'Sur les séries de Dirichlet', *Comptes Rendus*, 25 April 1910.

(2) 'Sur un théorème de M. Landau', *Comptes Rendus*, 22 Aug. 1910.

(3) 'A széttartó végtelen sorok elméletéhez', *Mathematikai és Természettudományi Értesitö*, vol. 29, 1911, pp. 719–726.

(4) 'Sur quelques généralisations d'un théorème de Weierstrass', *Comptes Rendus*, 21 Aug. 1911.

(5) 'Sur une propriété des racines des moyennes arithmétiques d'une série entière réelle', *Comptes Rendus*, 13 Oct. 1913.

(6) 'Vizsgálatok az absolut summabilis sorokról, alkalmazással a Dirichlet-és Fourier-sorokra', *Mathematikai és Természettudományi Értesitö*, vol. 32, 1914*.

M. Fekete and **G. Polya**

(1) 'Über ein Problem von Laguerre', *Rendiconti di Palermo*, vol. 34, 1912, pp. 89–120.

W. B. Ford

(1) 'On the relation between the sum-formulas of Hölder and Cesàro', *American Journal of Math.*, vol. 32, 1910, pp. 315–326.

M. Fujiwara

(1) 'On the convergence-abscissa of general Dirichlet's series', *Tôhoku Math. Journal*, vol. 6, 1914, pp. 140–142.

T. H. Gronwall

(1) 'Über die Laplacesche Reihe', *Math. Ann.*, vol. 74, 1913, pp. 213–270.

(2) 'Über die Summierbarkeit der Reihen von Laplace und Legendre', *Math. Ann.*, vol. 75, 1914, pp. 321–375.

A. Haar

(1) 'Über die Legendresche Reihe', *Rendiconti di Palermo*, vol. 32, 1911, pp. 132–142.

* We are unable to complete this reference at present.

J. Hadamard

(1) 'Sur les séries de Dirichlet', *Rendiconti di Palermo*, vol. 25, 1908, pp. 326–330, 395–396.

H. Hankel

(1) 'Die Euler'schen Integrale bei unbeschränkter Variabilität des Argumentes', *Zeitschrift für Math.*, Jahrgang 9, 1864, pp. 1–21.

G. H. Hardy

(1) 'On certain oscillating series', *Quarterly Journal*, vol. 38, 1907, pp. 269–288.

(2) 'The multiplication of conditionally convergent series', *Proc. Lond. Math. Soc.*, ser. 2, vol. 6, 1908, pp. 410–423.

(3) 'The application to Dirichlet's series of Borel's exponential method of summation', *Proc. Lond. Math. Soc.*, ser. 2, vol. 8, 1909, pp. 277–294.

(4) 'Theorems relating to the convergence and summability of slowly oscillating series', *Proc. Lond. Math. Soc.*, ser. 2, vol. 8, 1909, pp. 301–320.

(5) 'On a case of term-by-term integration of an infinite series', *Messenger of Math.*, vol. 39, 1910, pp. 136–139.

(6) 'Theorems connected with Maclaurin's test for the convergence of series', *Proc. Lond. Math. Soc.*, ser. 2, vol. 9, 1910, pp. 126–144.

(7) 'The multiplication of Dirichlet's series', *Proc. Lond. Math. Soc.*, ser. 2, vol. 10, 1911, pp. 396–405.

(8) 'An extension of a theorem on oscillating series', *Proc. Lond. Math. Soc.*, ser. 2, vol. 12, 1912, pp. 174–180.

(9) 'On the summability of Fourier's series', *Proc. Lond. Math. Soc.*, ser. 2, vol. 12, 1912, pp. 365–372.

(10) 'Note on Lambert's series', *Proc. Lond. Math. Soc.*, ser. 2, vol. 13, 1913, pp. 192–198.

G. H. Hardy and **S. Chapman**

(1) 'A general view of the theory of summable series', *Quarterly Journal*, vol. 42, 1911, pp. 181–216.

G. H. Hardy and **J. E. Littlewood**

(1) 'The relations between Borel's and Cesàro's methods of summation', *Proc. Lond. Math. Soc.*, ser. 2, vol. 11, 1911, pp. 1–16.

(2) 'Contributions to the arithmetic theory of series', *Proc. Lond. Math. Soc.*, ser. 2, vol. 11, 1912, pp. 411–478.

(3) 'Sur la série de Fourier d'une fonction à carre sommable', *Comptes Rendus*, 28 April 1913.

(4) 'Tauberian theorems concerning power series and Dirichlet's series whose coefficients are positive', *Proc. Lond. Math. Soc.*, ser. 2, vol. 13, 1913, pp. 174–191.

(5) 'Some theorems concerning Dirichlet's series', *Messenger of Math.*, vol. 43, 1914, pp. 134–147.

(6) 'New proofs of the prime-number theorem and similar theorems', *Quarterly Journal*, vol. 46, 1915 (unpublished).

(7) 'Contributions to the theory of the Riemann zeta-function and the theory of the distribution of primes', *Acta Math.* (unpublished).

A. Harnack

(1) 'Die allgemeinen Sätze über den Zusammenhang der Functionen einer reellen Variabeln mit ihren Ableitungen', *Math. Ann.*, vol. 23, 1884, pp. 244–284.

H. Heine

(1) 'Einige Anwendungen der Residuenrechnung von Cauchy', *Journal für Math.*, vol. 89, 1880, pp. 19–39.

O. Hölder

(1) 'Zur Theorie der trigonometrischen Reihen', *Math. Ann.*, vol. 24, 1884, pp. 181–211.

J. L. W. V. Jensen

(1) 'Om Rækkers Konvergens', *Tidsskrift for Math.*, ser. 5, vol. 2, 1884, pp. 63–72.

(2) 'Sur une généralisation d'un théorème de Cauchy', *Comptes Rendus*, 19 March 1888.

(3) 'Sur les fonctions convexes et les inégalités entre les valeurs moyennes', *Acta Math.*, vol. 30, 1899, pp. 175–193.

K. Knopp

(1) 'Grenzwerte von Reihen bei der Annäherung an die Konvergenzgrenze', *Dissertation*, Berlin, 1907.

(2) 'Multiplikation divergenter Reihen', *Sitzungsberichte der Berliner Math. Gesellschaft*, Jahrgang 7, 1907, pp. 1–12.

(3) 'Divergenzcharaktere gewisser Dirichletscher Reihen', *Acta Math.*, vol. 34, 1908, pp. 165–204.

(4) 'Nichtfortsetzbare Dirichletsche Reihen', *Math. Ann.*, vol. 69, 1909, pp. 284–288.

(5) 'Grenzwerte von Dirichletschen Reihen bei der Annäherung an die Konvergenzgrenze', *Journal für Math.*, vol. 138, 1910, pp. 109–132.

(6) 'Über die Abszisse der Grenzgeraden einer Dirichletschen Reihe', *Sitzungsberichte der Berliner Math. Gesellschaft*, Jahrgang 10, 1910, pp. 1–7.

(7) 'Bemerkung zu der vorstehenden Arbeit des Herrn I. Schur', *Math. Ann.*, vol. 74, 1913, pp. 459–461.

(8) 'Über Lambertsche Reihen', *Journal für Math.*, vol. 142, 1913, pp. 283–315.

T. Kojima

(1) 'On the convergence-abscissa of general Dirichlet's series', *Tôhoku Math. Journal*, vol. 6, 1914, pp. 134–139.

E. Landau

(1) 'Über einen Satz von Tschebyschef', *Math. Ann.*, vol. 61, 1905, pp. 527–550.

(2) 'Über die Grundlagen der Theorie der Fakultätenreihen', *Münchener Sitzungsberichte*, vol. 36, 1906, pp. 151–218.

(3) 'Über die Konvergenz einiger Klassen von unendlichen Reihen am Rande des Konvergenzgebietes', *Monatshefte für Math.*, vol. 18, 1907, pp. 8–28.

(4) 'Über die Multiplikation Dirichlet'scher Reihen', *Rendiconti di Palermo*, vol. 24, 1907, pp. 81–160.

(5) 'Beiträge zur analytischen Zahlentheorie', *Rendiconti di Palermo*, vol. 26, 1908, pp. 169–302.

(6) 'Neue Beiträge zur analytischen Zahlentheorie', *Rendiconti di Palermo*, vol. 27, 1909, pp. 46–58.

(7) 'Über das Konvergenzproblem der Dirichlet'schen Reihen', *Rendiconti di Palermo*, vol. 28, 1909, pp. 113–151.

(8) 'Über die Bedeutung einiger neuen Grenzwertsätze der Herren Hardy und Axer', *Prace matematyczno-fizyczne*, vol. 21, 1910, pp. 97–177.

(9) 'Über die Anzahl der Gitterpunkte in gewissen Bereichen', *Göttinger Nachrichten*, 1912, pp. 687–771.

(10) 'Sur les séries de Lambert', *Comptes Rendus*, 13 May 1913.

(11) 'Über einen Satz des Herrn Littlewood', *Rendiconti di Palermo*, vol. 35, 1913, pp. 265–276.

(12) 'Ein neues Konvergenzkriterium für Integrale', *Münchener Sitzungsberichte*, vol. 43, 1913, pp. 461–467.

(13) 'Die Identität des Cesàroschen und Hölderschen Grenzwertes für Integrale', *Leipziger Berichte*, vol. 65, 1913, pp. 131–138.

(14) 'Über die Nullstellen Dirichletscher Reihen', *Berliner Sitzungsberichte*, vol. 41, 1913, pp. 897–907.

(*See also* **H. Bohr** and **E. Landau**.)

E. Lasker

(1) 'Über Reihen auf der Convergenzgrenze', *Phil. Trans. Roy. Soc.* (A), vol. 196, 1901, pp. 431–477.

H. Lebesgue

(1) 'Recherches sur la convergence des séries de Fourier', *Math. Ann.*, vol. 61, 1905, pp. 251–280.

E. Lindelöf

(1) 'Quelques remarques sur la croissance de la fonction $\zeta(s)$', *Bull. des Sciences Math.*, ser. 2, vol. 32, 1908, pp. 341–356.

(2) 'Mémoire sur certaines inégalités dans la théorie des fonctions monogènes et sur certaines propriétés nouvelles de ces fonctions dans le voisinage d'un point singulier essentiel', *Acta Societatis Fennicae*, vol. 35, no. 7, 1909, pp. 1–35.

(*See also* **E. Phragmén** and **E. Lindelöf.**)

J. Liouville

(1) 'Sur quelques questions de géométrie et de mécanique, et sur un nouveau genre de calcul pour résoudre ces questions', *Journal de l'École Polytechnique*, vol. 13, cahier 21, 1832, pp. 1–69.

(2) 'Sur le calcul des différentielles à indices quelconques', *ibid.*, pp. 71–162.

J. E. Littlewood

(1) 'The converse of Abel's theorem on power series', *Proc. Lond. Math. Soc.*, ser. 2, vol. 9, 1910, pp. 434–448.

(2) 'Quelques conséquences de l'hypothèse que la fonction $\zeta(s)$ de Riemann n'a pas de zéros dans le demi-plan $R(s) > \frac{1}{2}$', *Comptes Rendus*, 29 Jan. 1912.

(*See also* **G. H. Hardy** and **J. E. Littlewood.**)

Hj. Mellin

(1) 'Die Dirichlet'sche Reihen, die zahlentheoretischen Funktionen, und die unendlichen Produkte von endlichen Geschlecht', *Acta Societatis Fennicae*, vol. 31, no. 2, 1902, pp. 1–48 (and *Acta Math.*, vol. 28, 1904, pp. 37–64).

F. Mertens

(1) 'Über die Multiplicationsregel für zwei unendliche Reihen', *Journal für Math.*, vol. 79, 1875, pp. 182–184.

G. Mittag-Leffler

(1) 'Sur un nouveau théorème dans la théorie des séries de Dirichlet', *Comptes Rendus*, 22 Feb. 1915.

C. N. Moore

(1) 'On the summability of the double Fourier's series of discontinuous functions', *Math. Ann.*, vol. 74, 1913, pp. 555–572.

N. E. Nörlund

(1) 'Sur les séries de facultés', *Comptes Rendus*, 4 May 1914.

(2) 'Sur les séries de facultés et les méthodes de sommation de Cesàro et de M. Borel', *Comptes Rendus*, 11 May 1914.

(3) 'Sur les séries de facultés', *Acta Math.*, vol. 37, 1915, pp. 327–387.

O. Perron

(1) 'Zur Theorie der Dirichletschen Reihen', *Journal für Math.*, vol. 134, 1908, pp. 95–143.

E. Phragmén

(1) 'Sur une extension d'un théorème classique de la théorie des fonctions', *Acta Math.*, vol. 28, 1904, pp. 351–368.

E. Phragmén and **E. Lindelöf**

(1) 'Sur une extension d'un principe classique de l'analyse et sur quelques propriétés des fonctions monogènes dans le voisinage d'un point singulier', *Acta Math.*, vol. 31, 1908, pp. 381–406.

S. Pincherle

(1) 'Alcune spigolature nel campo delle funzioni determinanti', *Atti del* IV *Congresso dei Matematici*, vol. 2, 1908, pp. 44–48.

M. Plancherel

(1) 'Sur la convergence et sur la sommation par les moyennes de Cesàro de $\lim_{z=\infty} \int_a^z f(x) \cos xy\, dx$', *Math. Ann.*, vol. 76, 1915, pp. 315–326.

A. Pringsheim

(1) 'Über Functionen, welche in gewissen Punkten endliche Differentialquotienten jeder endlichen Ordnung, aber keine Taylorsche Reihenentwickelung besitzen', *Math. Ann.*, vol. 44, 1894, pp. 41–56.

(2) 'Über das Verhalten von Potenzreihen auf dem Convergenzkreise', *Münchener Sitzungsberichte*, vol. 30, 1900, pp. 43–100.

(3) 'Über die Divergenz gewisser Potenzreihen an der Convergenzgrenze', *Münchener Sitzungsberichte*, vol. 31, 1901, pp. 505–524.

(4) 'Über den Divergenz-Charakter gewisser Potenzreihen an der Convergenzgrenze', *Acta Math.*, vol. 28, 1903, pp. 1–30.

B. Riemann

(1) 'Versuch einer allgemeinen Auffassung der Integration und Differentiation', *Werke*, pp. 331–344.

M. Riesz

(1) 'Sur les séries de Dirichlet', *Comptes Rendus*, 21 June 1909.

(2) 'Sur la sommation des séries de Dirichlet', *Comptes Rendus*, 5 July 1909.

(3) 'Sur les séries de Dirichlet et les séries entières', *Comptes Rendus*, 22 Nov. 1909.

(4) 'Une méthode de sommation équivalente à la méthode des moyennes arithmétiques', *Comptes Rendus*, 12 June 1911.

(5) 'Über einen Satz des Herrn Fatou', *Journal für Math.*, vol. 140, 1911, pp. 89–99.

(6) 'Sur la représentation analytique des fonctions définies par des séries de Dirichlet', *Acta Math.*, vol. 35, 1911, pp. 193–210.

(7) 'Sur un théorème de la moyenne et ses applications,' *Acta Szeged*, 1, 1922–23.

A. Rosenblatt

(1) 'Über die Multiplikation der unendlichen Reihen', *Bulletin de l'Académie des Sciences de Cracovie*, 1913, pp. 603–631.

(2) 'Über einen Satz des Herrn Hardy', *Jahresbericht der deutschen Math.-Vereinigung*, vol. 23, 1914, pp. 80–84.

W. Schnee

(1) 'Über irreguläre Potenzreihen und Dirichletsche Reihen', *Dissertation*, Berlin, 1908.

(2) 'Die Identität des Cesàroschen und Hölderschen Grenzwertes', *Math. Ann.*, vol. 67, 1908, pp. 110–125.

(3) 'Über Dirichlet'sche Reihen', *Rendiconti di Palermo*, vol. 27, 1909, pp. 87–116.

(4) 'Zum Konvergenzproblem der Dirichletschen Reihen', *Math. Ann.*, vol. 66, 1909, pp. 337–349.

(5) 'Über Mittelwertsformeln in der Theorie der Dirichletschen Reihen', *Wiener Sitzungsberichte*, vol. 118, 1909, pp. 1439–1522.

(6) 'Über die Koeffizientendarstellungsformel in der Theorie der Dirichletschen Reihen', *Göttinger Nachrichten*, 1910, pp. 1–42.

(7) 'Über den Zusammenhang zwischen den Summabilitätseigenschaften, Dirichletscher Reihen und ihrem funktionentheoretischen Charakter', *Acta Math.*, vol. 35, 1911, pp. 357–398.

I. Schur

(1) 'Über die Äquivalenz der Cesàroschen und Hölderschen Mittelwerte', *Math. Ann.*, vol. 74, 1913, pp. 447–458.

T. J. Stieltjes

(1) 'Sur une loi asymptotique dans la théorie des nombres', *Comptes Rendus*, 3 Aug. 1885.

(2) 'Note sur la multiplication de deux séries', *Nouvelles Annales*, ser. 3, vol. 6, 1887, pp. 210–215.

O. Stolz

(1) 'Beweis einiger Sätze über Potenzreihen', *Zeitschrift für Math.*, Jahrgang 20, 1875, pp. 369–376.

(2) 'Nachtrag zur XVI. Mittheilung im 20. Bande dieser Zeitschrift S. 369', *Zeitschrift für Math.*, Jahrgang 29, 1884, pp. 127–128.

A. Tauber

(1) ‘Ein Satz aus der Theorie der unendlichen Reihen’, *Monatshefte für Math.*, vol. 8, 1897, pp. 273–277.

O. Toeplitz

(1) ‘Über allgemeine lineare Mittelbildungen’, *Prace matematyczno-fizyczne*, vol. 22, 1911, pp. 113–119.

G. Vivanti

(1) ‘Sulle serie di potenze’, *Rivista di Mat.*, vol. 3, 1893, pp. 111–114.

S. Wigert

(1) ‘Sur quelques fonctions arithmétiques’, *Acta Math.*, vol. 37, 1914, pp. 113–140.

W. H. Young

(1) ‘Über eine Summationsmethode für die Fouriersche Reihe’, *Leipziger Berichte*, vol. 63, 1911, pp. 369–387.

(2) ‘On infinite integrals involving a generalisation of the sine and cosine functions’, *Quarterly Journal*, vol. 43, 1911, pp. 161–177.

(3) ‘Sur les séries de Fourier convergentes presque partout’, *Comptes Rendus*, 23 Dec. 1912.

(4) ‘On multiple Fourier series’, *Proc. Lond. Math. Soc.*, ser. 2, vol. 11, 1912, pp. 133–184.

Printed in the United States
By Bookmasters